T0140322

Editor-in-Chief

Prof. Janusz Kacprzyk
Systems Research Institute
Polish Academy of Sciences
ul. Newelska 6
01-447 Warsaw
Poland
E-mail: kacprzyk@ibspan.waw.pl

For further volumes:
http://www.springer.com/series/7092

Tokuro Matsuo and Ricardo Colomo-Palacios (Eds.)

Electronic Business and Marketing

New Trends on Its Process and Applications

Springer

Editors

Tokuro Matsuo
Advanced Institute of Industrial Technology
Tokyo
Japan

Ricardo Colomo-Palacios
Universidad Carlos III de Madrid
Leganés
Spain

ISSN 1860-949X   ISSN 1860-9503   (electronic)
ISBN 978-3-642-42827-2   ISBN 978-3-642-37932-1 (eBook)
DOI 10.1007/978-3-642-37932-1
Springer Heidelberg New York Dordrecht London

© Springer-Verlag Berlin Heidelberg 2013
Softcover re-print of the Hardcover 1st edition 2013
This work is subject to copyright. All rights are reserved by the Publisher, whether the whole or part of
the material is concerned, specifically the rights of translation, reprinting, reuse of illustrations, recitation,
broadcasting, reproduction on microfilms or in any other physical way, and transmission or information
storage and retrieval, electronic adaptation, computer software, or by similar or dissimilar methodology
now known or hereafter developed. Exempted from this legal reservation are brief excerpts in connection
with reviews or scholarly analysis or material supplied specifically for the purpose of being entered
and executed on a computer system, for exclusive use by the purchaser of the work. Duplication of
this publication or parts thereof is permitted only under the provisions of the Copyright Law of the
Publisher's location, in its current version, and permission for use must always be obtained from Springer.
Permissions for use may be obtained through RightsLink at the Copyright Clearance Center. Violations
are liable to prosecution under the respective Copyright Law.
The use of general descriptive names, registered names, trademarks, service marks, etc. in this publication
does not imply, even in the absence of a specific statement, that such names are exempt from the relevant
protective laws and regulations and therefore free for general use.
While the advice and information in this book are believed to be true and accurate at the date of publication, neither the authors nor the editors nor the publisher can accept any legal responsibility for any
errors or omissions that may be made. The publisher makes no warranty, express or implied, with respect
to the material contained herein.

Printed on acid-free paper

Springer is part of Springer Science+Business Media (www.springer.com)

Preface

This book includes theory and practice on e-business and marketing from an academic and professional viewpoint. In Chapter 1, editors introduce the topic by including an overview of emergent technologies in e-business and marketing. Chapter 2 presents a study devoted to investigate Customer Relationship Management adoption in Portugal among large Portuguese organizations. Chapter 3 is a contribution on a role of intermediate agents between suppliers and customers in semiconductor market. Chapter 4 presents Post-Via, a system that tries to unite on one platform the necessary components to perform traditional Customer Relationship Management functions and opinion mining techniques to provide services of direct marketing. Chapter 5 introduces a help desk support system to extract FAQ (Frequently Asked Questions) automatically, to retrieve FAQ for inquiry emails, and to show FAQ in users' inputting their inquiry e-mail. Chapter 6 is aimed to analyze the impact of advertising in one emergent market-place using system dynamics simulation. Chapter 7 introduces a system designed to apply caching techniques to Semantic Technology applications. Chapter 8 presents a study devoted to analyze the factors that determine the adoption of electronic commerce by consumers, by means of the examination of users' perceived compatibility with e-commerce and their prior experience of online purchasing. In Chapter 9, authors report an empirical study of an extended technology acceptance model (TAM) for online video services. Chapter 10 presents an initiative aimed to design a European professional certificate for the new role of social media networker. Chapter 11 describes a new type of context-aware advertising that employs context information obtained by means of ubiquitous sensors. Chapter 12 is a contribution that investigates customer characteristics of mail order industry to detect bad debt customers.

Dr. Matsuo and Dr. Colomo-Palacios are grateful to the authors and reviewers for their contribution to this work. Editors also acknowledge with their gratitude the editorial team of Springer-Verlag for their support during the preparation of the manuscript.

February 1, 2013

Dr. Tokuro Matsuo
Dr. Ricardo Colomo-Palacios

Acknowledgements

Dr. Tokuro Matsuo and Dr. Ricardo Colomo-Palacios are more than grateful to all authors for the outstanding contribution they present the high quality chapters and have made to the success of this publication.

Dr. Tokuro Matsuo expresses the deepest appreciation to his family and colleagues who have supported me in publishing this book.

Dr. Ricardo Colomo-Palacios would like to thank his (lovely and beautiful) wife Cristina, a constant source of inspiration, encouragement and endless love and Rodrigo, the charming child.

Contents

X   Contents

Towards New Generation E-Business and Marketing

Tokuro Matsuo[1] and Ricardo Colomo-Palacios[2]

[1] Advanced Institute of Industrial Technology 1-10-40, Higashi-Oi,
Shinagawa, Tokyo, Japan
[2] Universidad Carlos III de Madrid
Av. Universidad 30, Leganés, Spain

The ever-growing influence of the Internet has caused a paradigm shift in relationships between customers and companies [1]. For example, group buy marketplaces on the web broke down around 2005, but new types of group buy system like the Groupon appear after 2008 and become popular with the coupon-based business model and social network based system like the Facebook. And also, this is related with e-marketing because market makers can survey and investigate consumer's behavior, preference, trends of buying by collecting their information through AJAX-based systems (Javascript can send users operation information to website owner.).

In academic viewpoint, it is important to analyze the correlations between user preference/behavior and successful business case with system design and architecture. Further, to create new business model and system design with predicting user perspectives in new generation e-business and marketing, it is important to make an opportunity to present and discuss new research contributions.

Internet technology has significantly changed the ways in which firms collaborate and compete [2]. Companies are increasingly using Electronic Business applications such as electronic auctions, electronic catalogues, and customer relationship management applications to streamline their business processes along the entire supply chain [3]. Electronic Business can be defined as information systems to acquire, process, and transmit information for more effective decision-making, relative to competitive standards [4]. E-business is part of a wider economic context that is responsible for radical transformations in business and encompasses digital networks and communication infrastructure [5].

In this setup, Marketing functions have been transformed due to the Internet. Thus, [6] suggest these consequences in the Marketing scenario: de-intermediation, customer relationship management (CRM), mass customization, sales force automation, marketing decision support information, collaboration and coordination. Given that the Internet has enabled firms to reach out to global markets and has provided them with the opportunity to customize their strategies and offerings in an unprecedented way, the dynamics of international Internet marketing must be examined in deep [7].

Focusing on CRM systems several works highlighted the importance of the so-called Web 2.0 in its development [8–10] and there are several works devoted to draw the path from social media to social customer relationship management [11–16]. In this scenario, "Social CRM" is a novel concept that unites social media technology with customer relationship management [16]. With Social CRM communication

T. Matsuo & R. Colomo-Palacios (Eds.): Electronic Business and Marketing, SCI 484, pp. 1–3.
DOI: 10.1007/978-3-642-37932-1_1   © Springer-Verlag Berlin Heidelberg 2013

directed towards potential buyers can now be customized at an individual level through e-mails and social media [17]. This has changed contact management and reputation management in all kind of organizations around the globe.

Another aspect to take into account in CRM scene is mobility. Not in vain Mobile CRM has been tagged as the quiet explosion [18]. The emergence of mobile customer relationship management (mCRM) has made possible for the managers of the companies to access customer information from mobile devices without being sitting at their work place [19]. The future of mobile CRM is bright, not only for mobile workers using CRM applications but also for the digital clients who receive product and service offers via their phones [20]. Typical examples of mCRM are mobile marketing, mobile sales force automation and mobile field service, or mobile customer service [21]. As a result of its attractiveness there are many works in the literature devoted to study the phenomenon [21–27].

References

1. Colomo-Palacios, R., Varajão, J., Soto-Acosta, P. (eds.): Customer Relationship Management and the Social and Semantic Web. IGI Global (2011)
2. Pressey, A.D., Ashton, J.K.: The antitrust implications of electronic business-to-business marketplaces. Industrial Marketing Management 38(4), 468–476 (2009)
3. Bakker, E., Zheng, J., Knight, L., Harland, C.: Putting e-commerce adoption in a supply chain context. International Journal of Operations & Production Management 28(4), 313–330 (2008)
4. Jeffers, P.I., Muhanna, W.A., Nault, B.R.: Information Technology and Process Performance: An Empirical Investigation of the Interaction Between IT and Non-IT Resources*. Decision Sciences 39(4), 703–735 (2008)
5. Borges, M., Hoppen, N., Luce, F.B.: Information technology impact on market orientation in e-business. Journal of Business Research 62(9), 883–890 (2009)
6. Prasad, V.K., Ramamurthy, K., Naidu, G.M.: The Influence of Internet-Marketing Integration on Marketing Competencies and Export Performance. Journal of International Marketing 9(4), 82–110 (2001)
7. Eid, R., Elbeltagi, I., Zairi, M.: Making Business-to-Business International Internet Marketing Effective: A Study of Critical Factors Using a Case-Study Approach. Journal of International Marketing 14(4), 87–109 (2006)
8. González, D.P., González, P.S.: Interactions and Effects of CRM 2.0 in Public Administration. International Journal of Human Capital and Information Technology Professionals 3(1), 26–41 (2012)
9. García-Crespo, Á., Colomo-Palacios, R., Gómez-Berbís, J.M., Martín, F.P.: Customer Relationship Management in Social and Semantic Web Environments. International Journal of Customer Relationship Marketing and Management 1(2), 1–10 (2010)
10. García-Crespo, Á., Colomo-Palacios, R., Gómez-Berbís, J.M., Ruiz-Mezcua, B.: SEMO: a framework for customer social networks analysis based on semantics. Journal of Information Technology 25(2), 178–188 (2010)
11. Baird, C.H., Parasnis, G.: From social media to social customer relationship management. Strategy & Leadership 39(5), 30–37 (2011)
12. Greenberg, P.: The impact of CRM 2.0 on customer insight. Journal of Business & Industrial Marketing 25(6), 410–419 (2010)

13. Woodcock, N., Green, A., Starkey, M.: Social CRM as a business strategy. Journal of Database Marketing & Customer Strategy Management 18(1), 50–64 (2011)
14. Zhang, Z.: Customer knowledge management and the strategies of social software. Business Process Management Journal 17(1), 82–106 (2011)
15. Faase, R., Helms, R., Spruit, M.: Web 2.0 in the CRM domain: defining social CRM. International Journal of Electronic Customer Relationship Management 5(1), 1–22 (2011)
16. Baird, C.H., Parasnis, G.: From social media to Social CRM: reinventing the customer relationship. Strategy & Leadership 39(6), 27–34 (2011)
17. Nguyen, B., Mutum, D.S.: A review of customer relationship management: successes, advances, pitfalls and futures. Business Process Management Journal 18(3), 400–419 (2012)
18. Dickie, J.: Mobile CRM: The Quiet Explosion. CRM Magazine 15(6), 6–6 (2011)
19. Ranjan, J., Bhatnagar, V.: A holistic framework for mCRM – data mining perspective. Information Management & Computer Security 17(2), 151–165 (2009)
20. Goloenberg, B.: Mobile CRM: Nice to Have or a Business Essential? CRM Magazine 16(4), 8–8 (2012)
21. Schierholz, R., Kolbe, L.M., Brenner, W.: Mobilizing customer relationship management: A journey from strategy to system design. Business Process Management Journal 13(6), 830–852 (2007)
22. Sangle, P.S., Awasthi, P.: Consumer's expectations from mobile CRM services: a banking context. Business Process Management Journal 17(6), 898–918 (2011)
23. Sohn, C., Lee, D.-I., Lee, H.: The effects of mobile CRM activities on trust-based commitment. International Journal of Electronic Customer Relationship Management 5(2), 130–152 (2011)
24. Sinisalo, J., Karjaluoto, H.: Mobile Customer Relationship Management: a communication perspective. International Journal of Electronic Customer Relationship Management 1(3), 242–257 (2007)
25. Lee, D.-I., Sohn, C., Lee, H.: The role of satisfaction and trust in mobile CRM activities. International Journal of Electronic Customer Relationship Management 2(2), 101–119 (2008)
26. Smutkupt, P., Krairit, D., Khang, D.B.: Mobile marketing and consumer perceptions of brand equity. Asia Pacific Journal of Markcting and Logistics 24(4), 539–560 (2012)
27. Zheng, V.: The value proposition of adopting mCRM strategy in UK SMEs. Journal of Systems and Information Technology 13(2), 223–245 (2011)

Results of CRM Adoption in Large Companies in Portugal

João Varajão[1], Maria Manuela Cruz-Cunha[2], and Daniela Santana[3]

[1] University of Trás-os-Montes e Alto Douro, Portugal
[2] Polytechnic Institute of Cávado and Ave, Portugal
[3] Mestrado em TIC, University of Trás-os-Montes e Alto Douro, Portugal

Abstract. When a company implements a Customer Relationship Management system, there are several results that can be obtained as, for instance, the improvement of the information quality or the improvement of the customers' services. This paper presents the findings of a study undertaken among a sample of Portuguese large enterprises, identifying and discussing the main results of the adoption of CRM systems. The findings of this study help the academe and professional community to better understand the CRM adoption in large companies.

Keywords: Customer Relationship Management, CRM, Results, adoption, large companies, survey.

1 Introduction

Over the last 15 years, CRM (Customer Relationship Management has developed into an area of major relevance [14] , and despite significant interest from both academicians and practitioners, there remains a lack of agreement about what CRM is and how CRM strategy should be developed [24], as also whether it represents a huge investment with little measured payback [28] .

The term emerged in the information technology (IT) vendor community by the mid-nineties, often used to describe technology-based customer solutions, such as sales force automation [24], and is described as "information-enabled relationship marketing" in [30]. To Payne & Frow [24][25] there is a lack of a widely accepted and appropriate definition of CRM, which can contribute to the failure of a CRM project. This lack of a complete and integrated perspective, allows viewing this system as mere IT solution to gather clients, or a mere call centre or help desk, or customers' database. The authors state that the way CRM is defined is not only a semantic issue, as the picture that the organization makes of this system affects significantly the way an entire organization accepts and practices CRM. CRM is a strategy, not a solution, and can provide enormous competitive advantage if implemented in a co-operative environment [20][29], and the success of its implementation requires the committed involvement of senior management in promoting and supporting the concept of customer relationship management within the organization [29].

T. Matsuo & R. Colomo-Palacios (Eds.): *Electronic Business and Marketing*, SCI 484, pp. 5–13.
DOI: 10.1007/978-3-642-37932-1_2   © Springer-Verlag Berlin Heidelberg 2013

Broadly, CRM is a combination of a business and marketing strategy that integrates people, process, technology and all business activities, with the purpose of to attract and retain customers, provide analytical capabilities, reduce costs and increase profitability, by the consolidation of the principles of customer loyalty [37]; or as stated by Chen & Popovich [8], a technology that seeks to understand the company's customers.

Gartner Inc. reported that the market for CRM software achieved a growth in 2007 of 23.1 %, rising to a total of $8.1 billion, and worldwide CRM market revenue totalled $9.15 billion in 2008, a 12.5% increase from 2007 revenue [17]. In 2007 Forrester Research Inc. anticipated that revenues would continue to grow to $11 billion in 2010 [3-4]. However, on the other side, commercial market studies and literature refers the high failure rate of the CRM projects [13][14][15][18].

Given its major importance for business competitiveness over the last 15 years, literature is rich and many research projects have been and are being carried out to identify and understand the main motivations for CRM systems adoption, the difficulties occurred in its implementation, the obtained results, among many other aspects, aiming to improve the theory and practice of CRM planning and development. Together, these studies enable not only to understand the CRM field state-of-the-art, but also enable to understand their evolution over time.

In particular, as CRM is fostering the economy, the authors sought to understand the main results of CRM systems adoption, and undertook a study within a sample of large Portuguese enterprises.

The next section frameworks the main results associated to CRM adoption, the third section introduces the methodology followed in the study, section four discusses the results, and the last section closes the paper with some conclusions and future research directions.

2   Background

To operate in an e-Business environment, an organisation needs a good control of knowledge on its markets, customers, products and services, methods and processes, competitors, employee skills and its regulatory environment [27]. Findings of several studies, for instance [31][33][34], validate the belief that CRM is a critical success factor for business performance, and that the ability to retain its customers is essential to long term growth. Customer satisfaction and the quality of customer service help building trustworthy relationships with customers, promoting the increase of competitive advantage.

The purpose of this section is to highlight the several results of CRM systems implementation, and is based on a literature review. Among the several results, mentioned in literature [1][2][5][6][7][9][10][12][16][21][22][23][25][26][32][35] as being the most significant, we summarize the main results as:

- To increase the companies' knowledge with respect to its customers, in order to better understanding their needs and expectancies, to keep a customized relationship, aiming at new customers acquisition, improve customers' loyalty and their retention, and a fast response to their requests;
- To develop and offer customized products and services differentiated from products and services offered by the concurrence;
- To establish a close and fluid communication channel with actual and potential customers;
- To reduce the cost of sale and of after-sales service, increasing the effectiveness of a vendor in its role of acquiring new customers;
- To better align the company with the market;
- To contribute to improve internal processes within the organization: improved decision making process, sales efficiency; increased productivity and improved IT architecture;
- To aggregate value for the client, rationalizing the internal processes of new product development, allowing the company to know the needs not addressed and the characteristics of the product desired by segments of customers, and administration of the flow of demands, reducing customer's buying time and psychical and physical effort, optimizing after-sales service through the offer of specialized quality services.

3   Methodology

In order to understand several aspects of CRM adoption by large companies in Portugal, the authors carried out a study involving an online questionnaire, which call for response was sent via e-mail to a stratified random sample of 500 companies in the universe of the 1000 largest national companies in terms of turnover, according to the Portuguese National Institute of Statistics (INE).

The questionnaire was made available on an online platform in the period from 09/Feb/2009 to 11/May/2009, and comprehended four rounds of response. It was structured by thematic groups of questions and included several types of questions (multiple choice, free text). It was pre-tested and underwent an iterative process of content and clarity validation to get the final version. There were obtained 85 valid responses, corresponding to a 17% response rate. After the data collection it was carried out a descriptive statistical analysis.

The first group of questions in the questionnaire intended to characterize the companies participating in the study. The characterization of the companies is presented in Table 1. It is interesting to note that most companies have fewer than 2,000 employees and only 11% of the companies have more than 2,000 employees. In terms of turnover, companies are distributed across the intervals shown in Table 1. Note that a significant proportion of companies (44%) had a turnover of between 50 and 250 million Euros per year. The number of responses "Do not know / Do not answer" has a value of 12%.

Table 1. Characterization of participant companies

Number of employees	Percentage
1 to 200	26%
201 to 500	31%
501 to 2000	32%
2001 to 5000	8%
More than 5000	3%
Turnover (euros)	**Percentage**
Less than 5 000 000	0%
5 000 000 to 10 000 000	1%
10 000 001 to 50 000 000	29%
50 000 001 to 250 000 000	44%
250 000 001 to 500 000 000	5%
More than 500 000 000	9%
Do not know / do not answer	12%

From the set of 85 valid responses, only 21 companies use CRM systems (25%). Its characterization is presented in Table 2. This a quite small percentage when compared with the results of other studies of adoption; for instance a survey made in 2008 in Austria reported that 62% of the surveyed enterprises use a CRM system [36].

Table 2. Characterization of companies using CRM systems

Number of employees	Percentage
1 to 200	24%
201 to 500	32%
501 to 2000	29%
2001 to 5000	5%
More than 5000	10%
Turnover (euros)	**Percentage**
10 000 001 to 50 000 000	29%
50 000 001 to 250 000 000	42%
250 000 001 to 500 000 000	5%
More than 500 000 000	10%
Do not know / do not answer	14%

4   Main Results and Discussion

This section presents and discusses the main findings of the study, in what concern the perceived results of CRM systems adoption.

Within the study, participants were inquired about the relevance given to a set of results achieved with CRM implementation, in a scale from 0 (not important at all) to 5 (very important).

Table 3 presents the average and standard deviation obtained for each result included in the questionnaire and Figure 1 presents the ranking of the results perceived by large companies as a consequence of CRM systems adoption in terms of the average of the answers given by the participants that use CRM systems.

Table 3. Achieved results with CRM implementation

Achieved results with CRM implementation	Average	Std. Dev.
Improved quality of information	3.857	1.1952
Improved decision-making process	3.762	1.3002
Processes improvement	3.762	1.1792
Reduction in response time to requests	3.667	1.4944
Elimination of redundant activities	3.619	1.3220
Increased sales efficiency	3.619	1.5322
Increased productivity	3.476	1.2892
Improved customers' satisfaction	3.476	1.4703
Valuation of the company	3.333	1.4260
Improved customers' services	3.143	1.3148
Increased business results	3.095	1.6403
Customer loyalty	3.095	1.3750
Improved IT architecture	3.048	1.6875
More effective customer acquisition	3.000	1.5166
Creation of customized marketing messages	2.333	1.4944
Creation of customized products and services	2.238	1.6403

When analysing the results obtained with the adoption of CRM systems, we found that for most companies these systems contributed to an "Improved quality of information", reaching a very satisfactory result (3.86); the results "Improved decision-making process" and "Process improvement" were also highly valued items with the implementation of CRM systems. On the opposite side we find the "Creation of customized products and services", as a less important result of the implementation of these systems.

The findings suggest that the perceived results achieved with CRM implementation are in line with the motivations for CRM adoption identified by the authors in their study "Motivations for CRM adoption in large companies "[11], that is, it appears that the results of CRM adoption reflect the motivations of companies to adopt CRM systems. As referred in that study, an important motivation for CRM adoption is "Improving the quality of information" corresponding to the most significant result "Improved quality of information", as well as the "Improving the overall customer satisfaction" motivation, which was on top of the rank motivations, was reflected on the very significant outcome, "Reduction in the response time to requests" which reinforces the idea of the mission of a CRM system. "Process Improvement" being a very strong motivation, is also reflected in the results, assuming an even slight better position. Concerning the results of CRM adoption, which primarily focus on the customer, namely, "Improved customers' satisfaction" and "Improved customers' service" that were rated above the average, they correspond to motivations that were also highlighted as robust.

Fig. 1. Results of CRM adoption

5   Conclusions

The authors have undertaken a study directed to the largest Portuguese companies in order to perceive the results of companies regarding the utilization of a CRM system. Of the 85 firms that responded, 25% (21 companies) affirmed using CRM systems, which is a very low rate, given the importance that these systems may have in the development of the company competitiveness.

Within the set of results of CRM adoption, the better rated ones were related with improved quality of information and processes improvement.

One of the main limitations of the study is the reduced number of answers, only 21 of our inquired companies used CRM systems. A further research should be undertaken with a larger sample and an attempt to compare results with similar studies across the world should be tried.

References

1. Ahn, J.Y., Kim, S.K., Han, K.S.: On the design concepts for CRM system. Industrial Management & Data Systems 103(5), 324–331 (2003)
2. Alshawi, S., Missi, F., Irani, Z.: Organisational, technical and data quality factors in CRM adoption – SMEs perspective. Industrial Marketing Management 40(3), 376–383 (2010)
3. Band, W.: The Forrester wave: Enterprise CRM suites. Forrester Research, Inc. (2007), http://www.microsoft.com/presspass/itanalyst/docs/02052007ForrCRMSuites.pdf (retrieved August 20, 2012)
4. Band, W.: The Forrester wave: Enterprise CRM suites. Forrester Research, Inc. (2008), http://www.forrester.com/rb/Research/crm_best_practices_adoption/q/id/44179/t/2?src=46169pdf (retrieved August 20, 2012)
5. Becker, J.U., Greve, G., Albers, S.: The impact of technological and organ-izational implementation of CRM on customer acquisition, maintenance, and retention. International Journal of Research in Marketing 26(3), 207–215 (2009)
6. Bose, R.: Customer relationship management: key components for IT success. Industrial Management & Data Systems 102(2), 89–97 (2002)
7. Bull, C.: Customer Relationship Management (CRM) systems, intermediation and disintermediation: The case of INSG. International Journal of Information Man-agement 30(1), 94–97 (2010)
8. Chen, I.J., Popovich, K.: Understanding customer relationship management (CRM): People, process and technology. Business Process Management Journal 9(5), 672–688 (2003)
9. Chen, Q., Chen, H.-M.: Exploring the success factors of eCRM strategies in practice. The Journal of Database Marketing & Customer Strategy Management 11(4), 333–343 (2004)
10. Colomo-Palacios, R., Varajao, J., Soto-Acosta, P.: Customer Relationship Management and the Social and Semantic Web: Enabling Cliens Conexus. IGI Global, Hershey (2011)
11. Cruz-Cunha, M.M., Varajão, J., Santana, D.: Motivations for CRM Adoption in Large Companies in Portugal. Paper Presented at the Second International Conference on Business Sustainability, Póvoa de Varzim, Portugal (2011)
12. Ernst, H., Hoyer, W., Krafft, M., Krieger, K.: Customer relationship man-agement and company performance—the mediating role of new product performance. Journal of the Academy of Marketing Science 39(2), 290–306 (2010)

13. Foss, B., Stone, M., Ekinci, Y.: What makes for CRM system success — Or failure? Journal of Database Marketing & Customer Strategy Management 15, 68–78 (2008)
14. Frow, P., Payne, A.: Customer Relationship Management: A Strategic Per-spective. Journal of Business Market Management 3(1), 7–27 (2009)
15. Frow, P., Payne, A., Wilkinson, I.F., Young, L.: Customer management and CRM: addressing the dark side. Journal of Services Marketing 25(2), 79–89 (2011)
16. Garrido-Moreno, A., Padilla-Meléndez, A.: Analyzing the impact of knowl-edge management on CRM success: The mediating effects of organizational factors. International Journal of Information Management 31(5), 437–444 (2011)
17. Gartner. Gartner Says Worldwide CRM Market Grew 12.5 Percent in 2008. Gartner Inc. (2009), http://www.gartner.com/it/page.jsp?id=1074615 (retrieved August 14, 2012)
18. Kale, S.H.: CRM failure and the seven deadly sins. Marketing Management 13, 42–46 (2004)
19. Karakostas, B., Kardaras, D., Papathanassiou, E.: The state of CRM adoption by the financial services in the UK: an empirical investigation. Information & Management 42(6), 853–863 (2005)
20. Kotorov, R.: Customer relationship management: strategic lessons and future directions. Business Process Management Journal 9(5), 566–571 (2003)
21. Krasnikov, A., Jayachandran, S., Kumar, V.: The Impact of Customer Rela-tionship Management Implementation on Cost and Profit Efficiencies: Evidence from the U. S. Commercial Banking Industry. Journal of Marketing 73(6) (2009)
22. Osarenkhoe, A., Bennani, A.-E.: An exploratory study of implementation of customer relationship management strategy. Business Process Management Journal 13(1), 139–164 (2007)
23. Paguio, R.: CRM technology: can adoption increase service quality and per-ceived value in maintenance services? International Journal of Services Sciences 3(2-3), 250–268 (2010)
24. Payne, A., Frow, P.: A Strategic Framework for Customer Relationship Management. Journal of Marketing 69(4), 167–176 (2005)
25. Payne, A., Frow, P.: Customer Relationship Management: from Strategy to Implementation. Journal of Marketing Management 22(1), 135–168 (2006)
26. Peelen, E., van Montfort, K., Beltman, R., Klerkx, A.: An empirical study into the foundations of CRM success. Journal of Strategic Marketing 17(6), 453–471 (2009)
27. Plessis, M., Boon, J.A.: Knowledge management in eBusiness and cus-tomer relationship management: South African case study findings. International Journal of Information Management 24(1), 73–86 (2004)
28. Richards, K.A., Jones, E.: Customer relationship management: Finding value drivers. Industrial Marketing Management 37(2), 120–130 (2008)
29. Roberts, M.L., Liu, R.R., Hazard, K.: Strategy, technology and organisa-tional alignment: Key components of CRM success. The Journal of Database Marketing & Customer Strategy Management 12(4), 315–326 (2005)
30. Ryals, L., Payne, A.: Customer relationship management in financial services: towards information-enabled relationship marketing. Journal of Strategic Marketing 9(1), 3–27 (2001)
31. Saini, A., Grewal, R., Johnson, J.: Putting market-facing technology to work: Organizational drivers of CRM performance. Marketing Letters 21(4), 365–383 (2009)

32. Shamsuddoha, M., Nasir, T., Alamgir, M.: Determinants of Customer Rela-tionship Management (CRM): A Conceptual Analysis Fascicle of The Faculty of Economics and Public Administration 10(1) (2010)
33. Shin, I.: Adoption of Enterprise Application Software and Firm Performance. Small Business Economics 26(3), 241–256 (2006)
34. Sin, L.Y.M., Tse, A.C.B., Yim, F.H.K.: CRM: conceptualization and scale development. European Journal of Marketing 39(11/12), 1264–1290 (2005)
35. Smith, A.D.: Strategic Leveraging Total Quality and CRM Initiatives: Case Study of Service-Orientated Firms. Services Marketing Quarterly 32(1), 1–16 (2011)
36. Torggler, M.: The Functionality and Usage of CRM Systems. International Journal of Social Sciences 4(22), 164–172 (2009)
37. Wahab, S., Al-Momani, K., Noor, N.A.M.: The Relationship between E- Service Quality and Ease of Use On Customer Relationship Management (CRM) Per-formance: An Empirical Investigation In Jordan Mobile Phone Services. Journal of Internet Banking and Commerce 15(1), 1–15 (2010)

The Effectiveness of a New Product Coordinator in Market Access for a Semiconductor Venture

A Case Study of a Graphic Processing Unit LSI for Digital Pachinko

Akihiko Nagai[1], Hiroki Nakagawa[1], and Takayuki Ito[2]

[1] Department of Computer Science and Engineering, Nagoya Institute of Technology
Gokiso-cho, Showa-ku, Nagoya, Aichi, 466-8555 Japan
nagai.akihiko@nitech.ac.jp, nakagawa@itolab.nitech.ac.jp
[2] Master of Techno-Business Administration, Nagoya Institute of Technology
Gokiso-cho, Showa-ku, Nagoya, Aichi, 466-8555 Japan
ito.takayuki@nitech.ac.jp

Abstract. Venture companies that lack access to information cannot achieve market access. Even if a venture company is able to find a market, they may not be able to adequately get market needs. In the present case, a semiconductor venture company has been cooperated with the leading user company that linked by New Product Coordinator. This semiconductor venture company succeeded in developing an Application-Specific Standard Product (ASSP) and gaining market access. This paper shows that the New Product Coordinator (NPC) is valid for achieving market access for semiconductor venture companies.

1   Introduction

Venture companies face three problems when trying to gain market access. The first is to choose the accessing market. The appropriate market is one in which the venture can apply its core technology. The second is to correctly determine the potential market share. Venture companies must estimate the expected operating revenues. The third is to discover a market need. If the company's new product does not fill a need in the market, it will not be successful and the venture company will need to find a new value that addresses market needs. In addition, if another need is found through analysis of market need, creating a new value that addresses this need is also necessary.

The world semiconductor market has a fixated interest in the Application-Specific Standard Product (ASSP). An ASSP that addresses market needs will be highly successful and will provide a unique added value. Such an ASSP has been adopts by electronics user companies all over the world. In Japan, ASSPs are widespread mainly in medium-size electronics user companies. Thus, ASSPs offer an opportunity for venture companies to gain market access. Internationally, fabless semiconductor companies (such as Broadcom and Qualcomm) have grown significantly. Fabless semiconductor companies with sales of more than 10 billion yen are growing in Japan. These Japanese fabless semiconductor venture companies have resolved the three

T. Matsuo & R. Colomo-Palacios (Eds.): *Electronic Business and Marketing*, SCI 484, pp. 15–28.
DOI: 10.1007/978-3-642-37932-1_3   © Springer-Verlag Berlin Heidelberg 2013

problems discussed above. In the case that is the subject of this paper, a semiconductor distributor acting as a core agent provided the appropriate information to a fabless semiconductor company. The fabless semiconductor company and the semiconductor distributor cooperated to develop an ASSP. The fabless semiconductor company then successfully gained market access. To analyze this case study, clarifying the role of the core agent in resolving the three market access problems is important. In this paper, the core agent is a New Product Coordinator (NPC). An NPC is not merely a proponent of matching and problem solving. An NPC plays the central role of carrying out innovation.

This paper is presented as follows: The next section clarifies the position of this study through a review of previous research. Section three sets out the research purpose and research approach. Section four describes the problems faced by fabless semiconductor companies seeking market access. Market access requires a link between the core technology and market need. Section five describes the pachinko market, which is the target of this case study. Section six illustrates how the pachinko market was accessed by the venture. Section seven considers a method of solving the problem of quantity and function. Section eight proposes the New Product Coordinator (NPC). This paper concludes with an analysis of the results and a discussion of future research.

2   Previous Research on Inter-organizational Cooperation

Research on inter-organizational cooperation has been widely discussed from many viewpoints. Gray and Wood[6] analyzed the prerequisite of cooperation, the process of cooperation, and the results of cooperation from the multiple perspectives of resource dependence model, institutional economics, strategic management, social ecology, microeconomics, and political science. Bailey and Koney[4] analyzed cooperation from the perspectives of resource interdependency, social responsibility, and operational efficiency. Reitan[12] and Sasaki[13] analyzed cooperation on the basis of user requirement, professional, inter-organizational learning, microeconomics, resource dependence, and sociology of knowledge. Nagai and Tanabe[10] focused on partnerships for which one purpose of cooperation is to combine technology seeds and market needs. They determined that an enabler is necessary to carry this out, arguing that, without an enabler, it is difficult to solve problems among stakeholders.

Nagai and Tanabe[11] opined that, if cooperation allows for the use and sharing of specific information (secret information), cooperation will create a new value. However, specific information is not important; rather, it is necessary to extract essential important information from all the information available. Serious problems that often occur between stakeholders are also found in contract theory[7][8]. Even in this situation, if the behavior chosen by the stakeholder cannot be observed, moral hazards can occur and thus cause a serious incentive problem. In contract theory, a mediator is proposed as a means of resolving this problem. However, this mediator negotiates on behalf of the principal or agent. In our study, the NPC operates in a neutral setting and creates a new value for all stakeholders.

3   Purpose and Methods

This paper focuses on a semiconductor venture involving fabless semiconductor companies, specifically highlighting the following three important areas of market access:

— Viewpoint 1: Market demand (finding the appropriate market for the core technologies)
— Viewpoint 2: Potential sales volume (revenue to be obtained from the market)
— Viewpoint 3: Market need (to determine market need)

Market demand (viewpoint 1) is to find a market for the core technology.

Potential sales volume (viewpoint 2) is the revenue that the product of the semiconductor ventures can expect from the market.

Market need is the user need that the semiconductor ventures will fill in the new market.

In examining the semiconductor venture from these three viewpoints, one solution is to use the NPC. In this case, the semiconductor venture, in collaboration with the user company, linked the technology seeds with the user needs. This case study is based on the facts experienced by the author. Interviews with the parties provide an objective analysis. In the interviews, multiple parties were asked the same set of questions. In addition, this paper is supplemented with a collection of secondary data, such as websites and literature research.

4   Problems of Successful Market Access for the Semiconductor Venture

The problems faced by the semiconductor venture in accessing the Japanese semiconductor market are addressed in this section.

The leading Japanese semiconductor companies are unsatisfactory customers for the ASSP for the following three reasons: First, these companies have digital consumer electronics products, mobile terminals, and personal computer divisions in-house. In addition, in-house semiconductor demands have become the core of the semiconductor business. Furthermore, the user companies with a large demand for semiconductors have competing electronics products in the market. If major semiconductor companies develop an ASSP, their competitors will not adopt it.

Second, major semiconductor companies that do not have an in-house division for a particular segment of the business have developed strong ties with one leading user company. These strong ties are examples of inter-organizational cooperation as described by Granovetter[5]. Such inter-organization allows for the sharing and use of a large quantity of information. However, in such a relationship, information necessary for the development of an ASSP can only be obtained from one specific user company. Because of this, some of the information may be completely out of touch with market needs. Third, the leading Japanese user companies have a unique feature; these companies believe that product differentiation is crucial. They will try to integrate all functions in the system LSI (large scale integration). Top management of leading user companies in Japan has entrusted product development to the engineers. As a result, these engineers aim to produce high-performance and high-function products, and then develop a custom LSI. That is, a custom LSI is the core business of the system LSI of major semiconductor companies. These companies are not good at developing ASSPs, which provides a market access opportunity for semiconductor ventures. However, semiconductor ventures do not leverage this opportunity because they cannot resolve the following three problems:

— Problem 1: Semiconductor ventures cannot find an appropriate market in which to apply the core technology.
— Problem 2: Semiconductor ventures do not have adequate market sales channels (Japanese user companies have chosen reliable suppliers).
— Problem 3: Semiconductor ventures cannot determine market needs.

As stated above, this case study is analyzed based on these three problems. The semiconductor venture in this case successfully gained market access with a Graphic Processor Unit (GPU) LSI for digital pachinko.

5   The Successful Market Access of the Axell Semiconductor Venture with a GPU LSI for Digital Pachinko

5.1   Overview of the Pachinko Market

This section provides an overview of the pachinko market. From 1998 to 2007, the pachinko market was stable with a market size of 2.95 billion yen to 2.75 billion yen[17].

In 1998, LCD-screen digital pachinko became common in the pachinko market. Digital pachinko introduced a new customer segment and the number of pachinko players increased significantly from 20 million (1998) to 22 million (2002). The reason for this increase is the "fun of playing pachinko" that digital pachinko provides. Digital pachinko introduced a new reason to play pachinko, that is, for "fun," which drew in female users. Digital pachinko is changing the direction of pachinko. The "fun of playing" pachinko may have become more important than gambling. To further enhance the "fun" of pachinko," a high performance GPU LSI is needed. Pachinko companies have begun to recognize the importance of high-performance GPU LSIs. Many pachinko companies tried to adopt the YGV622-S (YAMAHA LSI)[18]. However, the YGV622-S cannot change animation to computer graphics (CG). In addition, digital pachinko animation needs to repeat spinning, reduction, and expansion, areas where the YGV622-S's performance is low. However, the only GPU LSI that could be used for digital pachinko was the YGV622-S. Many pachinko companies considered the possibility of developing a GPU LSI, but abandoned it because custom LSI development requires a professional engineer and is costly (most likely more than 50 million yen).

Table 1. Changes in the market size of pachinko

Year	Market size (billion yen)	Population of participators (thousands)
1998	28,426,000	23,100
1999	28,057,000	19,800
2000	28,469,000	18,600
2001	28,697,000	20,200
2002	27,807,000	19,300
2003	29,225,000	21,700
2004	29,634,000	17,400
2005	29,486,000	17,900
2006	28,749,000	17,100
2007	27,455,000	16,600
2008	22,980,000	14,500
2009	21,716,000	15,800

Reference: Based on White Paper of Leisure (Japan Productivity Center: 1998-2009).

5.2 Leading Pachinko Company X

Company X is a leading pachinko company listed on the Tokyo stock exchange. Approximately 90% of all sales involve the manufacture and sale of pachinko machines. In 1998, pachinko had reached its height and had become a social problem. Pachinko was evaluated for permit approval by the Security Electronics and Communications Technology Association (SECTA)[14]. SECTA determined pachinko to be a significant social problem and thus obtained control of pachinko authorization. The pachinko industry was significantly impacted as a result. Company X's sales dropped by as much as 50 percent (approximately 700 billion yen). In response, pachinko companies are trying to rehabilitate pachinko's negative societal image with this new "fun of playing pachinko" theme. Company X, as a leader in the pachinko industry, has decided to make the transition to digital pachinko and thus is considering getting a high-performance GPU LSI. Company X does not have engineers knowledgeable in the design of LSIs. For this reason, Company X consulted a semiconductor distributor (Company Y) that has an LSI design department, which ultimately resulted in the cooperative semiconductor venture Axell, which has image processing technology as its core technology. Company Y proposed the cooperative development of these three companies to Company X. Company Y concluded that the core technology of Axell was capable of CG. However, the development of a custom GPU LSI was immediately scrapped for two reasons: The first was that the development of the GPU LSI would incur enormous development costs (probably 100 million yen or more). The second was that Axell was such a small venture. As a result, Company X might not be able to obtain sufficient support. Thus, the project was dissolved. However, Company Y proposed developing a GPU as an ASSP through the Axell semiconductor venture and Company X. Company Y investigated this idea by studying its feasibility with market users through interviews with 20 pachinko companies. Company Y has strong ties with many pachinko companies through its daily operations. This interview survey revealed that many pachinko companies would like to have CG, but the YGV622-S is not CG capable. Many pachinko companies were dissatisfied with the YGV622-S. Company Y considered this to be a new business opportunity. Company X discontinued the prospect of a custom GPU LSI, but understood that digital pachinko requires a new GPU LSI. If Company X accepted Company Y's venture proposal, Company X could get a new GPU LSI. On the other hand, if the Axell venture accepted this proposal, it can get Company X as a GPU LSI user. Axell and Company X agreed.

6   Market Access of Semiconductor Venture Axell

This section describes how the new GPU LSI was adopted for digital pachinko.

6.1   Semiconductor Venture Axell Developed a New GPU LSI for Digital Pachinko

Founded by Mr. Sasaki, the fabless semiconductor venture Axell was established in 1996. Currently, the GPU LSI(product AG-3) has been adopted by more than 40 pachinko companies. Axell introduced product AG-1 in 1998, developed product AG-2 in 2002, and developed product AG-3 in 2006. In 1998, Axell was a small venture company with few employees. Currently, Axell is listed on the Tokyo Stock Exchange.

The sales volume for the GPU LSI was 28 million pieces in 1999, 130 million pieces in 2002, and 262 million pieces in 2009, and it has continued to increase every year. Table 2 shows the trends of the GPU LSI market share. The year 2009 has accounted for 60% of the total digital pachinko production [3].

Table 2. Trends in market share of the GPU LSI

Year	1998	2002	2006	2009
GPU LSI	AG-1	AG-2	AG-3	AG-3
Sales volume of GPU LSI (Thousand pieces)	300	1,300	1,800	2,620
Production of digital pachinko (Thousand units)	3,133	4,676	5,484	4,097
Market share	10%	30%	47%	58.1%

Reference: Based on Investors Relations (Axell) and Yano Research Institute, "Pachinko Kanren Kigyou no Doukou to Market Share" (1998-2009).

Axell is leveraging the resources of universities to develop core technologies and collaborated with the University of Tsukuba from 1998 until 2006[16]. GPU LSIs (product AG-2 and product AG-3) adopted these core technologies. Venture companies can resolve human resource problems by using university resources. The core technology of Axell is the Adaptive Orthogonal Transform (RACP) and the Recurrent

AC Component Prediction (AOT). Specifically, AOT achieves smooth animation replay. This technology resolves both the mosquito noise and the block noise that occur with outline information. This phenomenon appears predominately in artificial images such as CG and animation, as opposed to natural pictures. RACP minimizes the diffusion of non-stationary portions that are used for multi-resolution decomposition on compression distortion to diffuse throughout the block. Axell has eight patent applications on these technologies (including the related patent). Recurrent-ACP-processed image code (RAPIC) was developed based on RACP technology. RAPIC is data compression technology that has far-reaching effects on the replay of CG and animation in particular. Company X saw the potential need of those who wanted new data compression technology. Potential needs are not concrete needs, such as "want to produce CG." The data size of CG is larger than animation. RAPIC has been developed to solve this future problem. Finding potential needs is an important integrant in product development.

6.2   Finding the Appropriate Market for the Core Technologies

Company Y found that the core technology of Axell is appropriate for digital pachinko. Axell was judged to have the opportunity for market access of the GPU LSI based on a lot of information about the semiconductor for three reasons. The first is that the pachinko market has a relatively large demand. Figure 1 is the result of a production volume survey of pachinko by the Yano Research Institute. According to the survey, the production volume of pachinko in 1998 was 3.13 million units. Most of this was for digital pachinko and this provides a sufficient market volume.

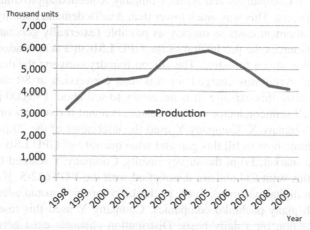

Reference: Based on the Yano Research Institute, "Pachinko Kanren Kigyou no Doukou to Market Share" (1998-2009).

Fig. 1. Production volume of pachinko

Second is that the GPU LSI will achieve higher and higher performance in the future. In the semiconductor market, semiconductor price is declining daily.

However, if digital pachinko adopts CG, high-definition image quality will be required. As a result, the GPU LSI will be on the pricey side.

Third, major semiconductor companies do not view the pachinko market positively. The leading semiconductor companies do not want to get involved in the gambling industry. The major semiconductor companies believe that their involvement will lead to a loss of partnership with leading user companies.

7   Building of Sales Channels and Meeting Market Needs

In the previous section, this paper described how Company Y found the applications market of Axell's core technologies and then linked Axell with a major user company to hit market needs. This section shows how sales channels were built and market needs met.

7.1   Method of Building Sales Channels (Problem of Minimum Need Sales Volume)

One of the expectations of a particular user company of a fabless semiconductor venture is to be the first user of the GPU LSI. However, the quantity purchased by a particular user company often falls short of the expectation of the fabless semiconductor venture. In this case study, the purchase quantity by the leading pachinko company was less than the quantity needed to be sold by Axell. As a result, a quantity problem occurred between Company X and Axell. Company X needed approximately 50 thousand GPU LSI pieces. This was much lower than Axell's demand quantity. Axell must recover its development costs as quickly as possible (externally generated costs) because Axell outsources the production of the GPU LSIs to a semiconductor company (referred to as the silicon foundry). The silicon foundry converts the design data into production data. Axell was charged for this data conversion at 80 million yen by Fujitsu. To amortize this expense, it is necessary to sell about 135,000 pieces (Suzuki[15]).The Axell semiconductor venture business cannot thrive only on the quantity purchased by Company X. Company Y used the interviews of the 20 pachinko companies to determine how to fill this gap and what quantity of GPU LSIs could be sold in the pachinko market. From the survey results, Company Y learned that many pachinko companies want CG and are dissatisfied with the YGV622-S. If the GPU LSI achieved CG in digital pachinko, many pachinko companies would adopt it. With its strong ties with many pachinko companies, Company Y used this research and exchanged information on a daily basis. Distribution channels exist between 80% or more of the semiconductor distributors in the Japanese semiconductor market. A leading semiconductor distributor such as Company Y has strong ties with the leading user companies in all segments (product business areas).

7.2   Fulfilling Market Needs (The Problem of Meeting "Must Have" Needs)

The fabless semiconductor venture also undertakes the role of providing market needs. This market need can be classified into the following three types; Must have needs (essential); nice to have needs (convenience); and specific needs (unnecessary).

The first, "must have," is an essential need that must be fulfilled.

"Nice to have" is a useful need. These needs are not so important.

"Specific" is a need unique to a particular user company. These needs are unnecessary needs. In the collaboration between Company X and the fabless semiconductor venture Axell, Company Y provides all needs (not only "must have," but also "nice to have" and "specific") to Axell. Axell organizes these needs and must extract the essential needs. On the other hand, the "nice to have" and "specific" needs are eliminated as much as possible because these factors drive up development costs. Company X delineated the following seven needs:

A. Implementation of CG
B. Improved performance of sprites (smooth spin, zoom in, and zoom out)
C. Minimization of the effects of static electricity
D. High compression rate algorithm development
E. Built-in D-RAM (Built-in cache memory)
F. Realization of 3D (three-dimensional animation)
G. Adoption of ball grid array (BGA) package (high-density package)

All of these needs are "must have" needs for Company X, but are not market needs. The "must have" contains this information. However, Axell cannot organize according to this information. If Axell cannot develop the GPU LSI, a new value will not be created, not only for the fabless semiconductor venture Axell, but also for Company X. Table 3 summarizes the problem between Company X and Axell.

Table 3. Different positions of Company X and Axell

	Leading specific pachinko company Company X	Fabless semiconductor venture Axell
Demand	Minimum purchase quantity is 50 thousand pieces a year	Minimum demand quantity is 13.5 thousand pieces in order to pay development costs to Fujitsu
Needs	X has seven needs for the GPU LSI and has evaluated them all as "must have"	GPU LSI should achieve only "must have" needs, but Axell does not know these "must have" needs

If Company X and Axell cannot develop the GPU LSI, they will miss a great opportunity. That is, Axell will miss an opportunity to access the market and Company X will not get CG in its digital pachinko.

Company Y resolved this problem as follows: Once again, Company Y used a network linked by strong ties. Company Y interviewed the 20 leading pachinko companies again. In this survey, the needs of Company X were presented as draft performances of the GPU LSI and these performances confirmed its usefulness. The "must have" (essential needs), "nice to have" (convenience needs) and "specific" (unnecessary needs) were manifested in this survey. Axell made its decision on the performance of the GPU LSI based on the categorization data from Company Y. As shown in Table 4, product AG-2 met the following needs:

A. Implementation of CG
B. Improved performance of CSS sprites (smooth spin, zoom in, and zoom out)
C. Minimization of the effects of static electricity
D. High compression rate algorithm development

In addition, product AG-3 met need E and need F:

E. Built-in D-RAM (Built-in cache memory)
F. Realization of 3D (three-dimensional animation)

Needs E and F were evaluated as "nice to have" (convenience needs) and meeting these needs was delayed until the development of product AG-3. The engineers of the pachinko companies, who took part in Company Y's survey, evaluated these needs as "useful, but not needed immediately." As a result, Company X and Axell agreed on the needs and the GPU LSI was developed.

Table 4. Performance comparison of GPU LSI

	Needs	AG-2[1]	AG-3[2]
Type of drawing	B	Sprite	Sprite
Type of drawing buffer		Double frame buffer	Double frame buffer
Maximum rendering performance	A	100 million dots / sec	400 million dots / sec
Graphics	A	CG (2D)	CG (3D)
VRAM	E	None	60 Mbit
Data compression technology	D	RAPIC	RAPIC

Reference: Based on data sheet.

One of the features proposed by Company Y provided a new solution for pachinko companies. In the future, the competitive power of digital pachinko will be determined by "fun." Leading specific pachinko company Company X has appreciated the advantage of being able to produce interesting content, rather than dealing with the problems that competitor companies face obtaining the new GPU LSI. That is, Company Y has grasped the value of the GPU LSI. Thus, the GPU LSI has been adopted by many pachinko companies and has become the de facto standard in the market.

8   Effectiveness of the New Product Coordinator (NPC) with the Semiconductor Distributor

8.1   The Active Process of the NPC

In this section, this paper proposes a new product coordinator based on the case study. Figure 2 shows the relationship between the semiconductor distributor (NPC) with the semiconductor venture and the particular user company. The semiconductor venture will be able to access the market through the support of the NPC in the following step: The NPC collects information on the semiconductor industry, such as technology seeds of semiconductor ventures and needs of user companies. In other words, the NPC itself is the collective intelligence of the semiconductor industry. Collective intelligence is shared or group intelligence that emerges from the collaboration and competition of many individuals. The NPC performs interactive communication between stakeholders in the following steps:

STEP 1: Search of user companies.
The user company explains the performance need to the semiconductor and requests information.

STEP 2: Proposal by NPC.
Using the collective intelligence, the NPC seeks an LSI that complies with the needs of the user company and delivers this information to the user company. If the LSI is not found, the NPC searches the core technology of semiconductor ventures. The NPC finds the core technology that can achieve the needs of the user and proposes the development of the LSI to the user companies.

STEP 3: Consideration by user companies.
The user companies evaluate the core technology proposed by the NPC and, if a problem is not found, consider development costs. The user company adapts to the terms and makes a decision to develop a custom LSI (user dedicated LSI). The semiconductor venture is formed to undertake the custom LSI business (the NPC links the user company and the semiconductor venture).

If the LSI development does not comply with the required conditions, the NPC will consider whether it is possible to do business via an ASSP by evaluating the feasibility of the ASSP business with users in the market.

STEP 4: Finding the market of the core technologies.
If, through the survey of user companies, it becomes apparent that the ASSP business is not feasible, the NPC's activity is finished (the NPC determines that there is no value for the business in the market). If the NPC finds that there is a business opportunity for the ASSP, the NPC will proceed to the next step.

STEP 5: Linking of core technology and user needs.
The NPC proposes the development of the ASSP to the semiconductor venture (the core technology owner) and a specific user company (ASSP customer) (to establish an application market for the core technology).

STEP 6: Resolving the problem of required sales quantity.
Quantity problems arise if the purchase quantity of the user company is less than the requirements of the semiconductor venture. If this problem occurs, the NPC confirms the purchase intention to the user companies of STEP 4. If many user companies express an intention to purchase the ASSP and the total purchase quantity of these companies can exceed the required quantity of the semiconductor venture, the NPC contracts a sales quantity with the semiconductor venture. When the total purchase quantity of these companies cannot exceed the required quantity of the semiconductor venture, the NPC proposes discontinuing the development of the ASSP to the particular user company and semiconductor venture (the NPC will give up the new ASSP business).

STEP 7: Understanding the essential needs.
All of the needs of a particular user company are not "must have" needs (essential needs). These needs include "nice to have" (convenience needs) and/or "specific" (unnecessary needs). The NPC confirms with the user companies of STEP 4 and STEP 6 whether the ASSP will successfully meet the needs of the particular user company. The NPC classifies the needs based on the results of this research and proposes that the ASSP's performance is required to meet only the "must have" needs to the particular user company and semiconductor venture. At the same time, the NPC indicates the new value created by this ASSP to these companies (getting essential needs).

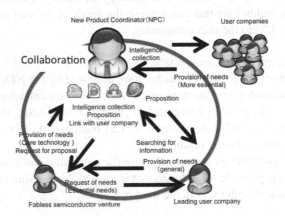

Fig. 2. New combination of New Product Coordinator

8.2   Capability Requirement for the NPC

The NPC proposed in this paper supports the market access of the semiconductor venture with limited resources. To do so, the NPC must have the following three capabilities:

Capability 1: Having strong ties.
It is necessary to have strong ties with the user and related companies in all business segments. In addition, the ties should be built upon a strong, trustful relationship.

Capability 2: Access to collective intelligence.
The NPC has been collecting all the information of the semiconductor industry. In other words, the NPC knows the core technology of almost all semiconductor ventures and also has information on many business segments. In addition, the NPC's strong ties enable it to handle marketing issues.

Capability 3: To propose a solution to the problem.
The NPC proposes a new business and resolves problems that arise during its implementation. In addition, the NPC creates new value with this new business.

In general, semiconductor distributors exhibit capabilities 1 and 2 in their daily sales activities. However, most semiconductor distributors do not perform capability 3. In order for semiconductor distributors to create new business like the NPC, they need capability 3.

9   Conclusion

This paper clarifies the problem of market access for semiconductor ventures from the three viewpoints listed below and proposes the semiconductor distributor role of the New Product Coordinator to support these companies.

— Viewpoint 1: Market demand (finding the appropriate market for the core technologies)
— Viewpoint 2: Potential sales volume (revenue to be obtained from the market)
— Viewpoint 3: Market needs (to determine market needs)

Furthermore, this paper shows that the NPC must exhibit three areas of capability in order to carry out this role, the first two of which semiconductor distributors already possess.

— Capability 1: Having strong ties
— Capability 2: Having collective intelligence
— Capability 3: To propose a solution to the problem

The focus of our future work will be to demonstrate the effectiveness of the NPC in many more cases.

References

1. AG-2, http://www.axell.co.jp/jp/products/catalog/ag-2.pdf
2. AG-3, http://www.axell.co.jp/jp/products/list/ag3.html
3. Axell, http://www.axell.co.jp/index.html

4. Bailey, D., Koney, K.: Strategic Alliances among Health and Human Service Organizations. Sage Publications, Inc., CA (2000)
5. Granovetter, M.: The Strength of Weak Ties. American Journal of Sociology 78(6), 1360–1380 (1973)
6. Gray, B., Wood, D.J.: Collaborative Alliances: Moving from Practice to Theory. Journal of Applied Behavioral Science 27(1), 3–32 (1991)
7. Ito, H.: A Course in Contract Theory. Yuikaku Publishing (2004) (in Japanese)
8. Ito, H.: Contract Theory: Path toward the Third Theory of Microeconomics. Journal of Economic Literature (JEL) classification numbers: B 21, D 86 49(2), 52–62 (2007)
9. Japan Productivity Center, White Paper of Leisure (1998-2009)
10. Nagai, A., Tanabe, K.: The Enabler Role of Japan's Semiconductor Distributors-Linking Technology with Market Needs. Journal of Japan Society for Intellectual Production 6(1), 23–33 (2007) (in Japanese)
11. Nagai, A., Tanabe, K.: Sharing and Using of Secret Information for Collaborative Innovation: Case Study on Development of ASSPs. Development Engineering 30(2), 133–142 (2011) (in Japanese)
12. Reitan, C.T.: Theories of Inter-organizational Relations in the Human Services. Social Service Review 72(3), 285–300 (1998)
13. Sasaki, T.: Sosiki-kan Kankei-ron no Kadai to Tenkai. In: Akaoka, I. (ed.) Keiei Senryaku to Sosiki-kan Teikei no Kouzu, CHUOKEIZAI-SHA, pp. 29–45 (2005)
14. Security Electronics and Communications Technology Association (SECTA), http://www.hotsukyo.or.jp/index.html
15. Suzuki, M.: Yokuwakaru Handoutai-LSI no Dekirumade. In: THE NIKKAN KOGYO SHIMBUN (2001)
16. University of Tsukuba Chaos and Computer Amusement Oriented Systems Laboratory, http://www.chaos.cs.tsukuba.ac.jp/index.html
17. Yano Research Institute, Pachinko Kanren Kigyou no Doukou to Market Share (1998-2009)
18. YAMAHA, YGV622-S Data Sheet, YAMAHA LSI, http://ms-n.org/DataSheets/Yamaha/htm

POST-VIA: Develop Individualized Marketing Strategies for Tourists

Antonio Cabanas-Abascal[1], Alejandro Rodríguez-González[2],
Cristina Casado-Lumbreras[3], Joaquín Fernández-González[4],
and Diego Jiménez-López[4]

[1] Universidad Carlos III de Madrid, Spain
[2] Centro de Biotecnología y Genómica de Plantas, Universidad Politécnica de Madrid, Spain
[3] Universidad Internacional de La Rioja, Spain
[4] Egeo IT, Spain

Abstract. POST-VIA is an information system whose main objective is develops tools to manage direct marketing strategies in the tourism sector. POST-VIA can be considered as software able to collect information about the travel experience for tourists and convert this information into knowledge. The system offers DMOs a management component of communication and interaction with the customer based on a highly accurate perception of it, allowing individualized marketing campaigns (Social Semantic CRM). Social Semantic CRM component incorporates several techniques to achieve this aim, among others, opinion mining, recommendation systems, and digital footprint. As a basic differential, POST-VIA platform is not limited to rely on the goodwill of tourists (often controversial and always random) to complete the valuable data of subjective perception, it offers an attractive product catalog and services compelling enough to take the time and the interest to collaborate.

1   Introduction

A key element of a successful tourism industry is the ability to recognize and deal with change across a wide range of key factors and the way they interact [19]. For example, investments in technology may often lead to improvements and enhancement of services, while the availability of technology may also affect the ways that marketing is undertaken [7]. Not in vain, technology investments are the anchor of mainstream process innovation, sometimes in combination with reengineered layouts for manual work operations [30].

   The coming decade and a half will see major shifts in the leisure and tourism environment reflecting changing consumer values, political forces, environmental changes and the explosive growth of information and communication technology [19]. In this environment, Kenteris, Gavalas and Economou [37] stated that the convergence of IT and communications technologies and the rapid evolution of the Internet have been some of the most influential factors in tourism that have changed travelers' behavior. As result of this, the tourism industry is demanding an ever-increasing level of value-added services in technologically complete environments, which are integrated and

T. Matsuo & R. Colomo-Palacios (Eds.): *Electronic Business and Marketing*, SCI 484, pp. 29–42.
DOI: 10.1007/978-3-642-37932-1_4   © Springer-Verlag Berlin Heidelberg 2013

highly dynamic [23]. According to these authors, administrative and corporate bodies in the tourism industry now have to focus on the development of new infrastructures, providing citizens with access to cultural content and tourism services.

Increasingly, ICTs play a critical role for the competitiveness of tourism organizations and destinations as well as for the entire industry as a whole [57]. For example, the potential to be gained from investment in effective online services, including higher bookings, return custom and the development of loyalty, is very substantial [27]. It is quite clear from the "U.S. Online Travel Overview, Eleventh Edition" report [49] that said online travel penetration in leisure/unmanaged business travel, measured in percentage of gross bookings, stood at around 39% in the EE.UU, 38% in Western Europe, 23% in Asia-Pacific, 18% in Latin America. Furthermore online travel revenues accounted for 59% of total travel in the EE.UU, 43% in Europe and 9% in Asia-Pacific. Consequently, ICTs have also changed radically the efficiency and effectiveness of tourism organizations, the way that businesses are conducted in the marketplace, as well as how consumers interact with organizations [5], [6].

At the same time, tourist behavior patterns have changed. The needs of consumers, who are increasingly less loyal, take more frequent vacations of shorter duration, and take less time between choosing and consuming a tourism product [60]. However, there is a clear trend of autonomous behavior of consumers, the tourists are able to contract different tourist services technologies based on self-service. There is an apparent dichotomy between, on the one hand building strong and lasting customer relationships while on the other encouraging consumer autonomy by developing more self-service technologies [53]. For this reason CRM is widely used in the tourism industry, with loyalty programs keeping customers returning and travel websites yielding a large volume of e-transactions [59].

Developments in search engines, carrying capacity and speed of networks have influenced the number of travellers around the world that use technologies for planning and experiencing their travels [15]. However, the expansion of Internet and the advent of Web 2.0 has caused a huge increase in available information, this makes more difficult the appropriate choice of destination, this task becomes tedious work of searching. Considering that four out of ten international visitors (38%) choose their destination based on friends & relatives' recommendation in 2011, and two out of ten choose this based on information on the web (about tourism, October 2011), is necessary to pay attention to information overload problem.

Currently there are several technologies that have sought a solution to this problem. Due to the large amount of information present in the world today, technologies are required which filter all the available data, leaving us with only the information which is valuable to us [45]. Thus online readers are in need of tools to help them cope with the mass of content available on the World-Wide Web [2]. Recommender systems have proven to be an important response to the information overload problem, by providing users with more proactive and personalized information services [44]. That is why, to be able to deal with the continuous growth of the WWW (in size, languages and formats), we need to exploit other information. This is where the Semantic Web comes in [3]. Besides, sentiment-analysis and opinion-mining systems also have an important potential role as enabling technologies for other systems, among others, recommendation systems and information extraction systems [46].

Therefore the amount of information available on, mainly as opinions, makes the task of making a decision tedious and difficult. Due to great importance that user give to these, POST-VIA tries to unite on one platform the necessary component to perform traditional CRM functions and opinion mining techniques to provide services of direct marketing using web semantic components and recommender systems. Consequently the amount of information available on, mainly as opinions, makes the task of making a decision tedious and difficult. Due to great importance that user give to these, POST-VIA tries to unite on one platform the necessary component to perform traditional CRM functions and opinion mining techniques to provide services of direct marketing using web semantic components and recommender systems. The system also offers different services to obtain the cooperation of user; these components get the indispensable social content for the correct operation.

The remainder of the chapter is structured as follows. First we make a brief review of the state of the art about innovative technologies used in the project: semantic web, opinion mining and recommender systems. Then we describe the architecture of the platform as well as some example of use. The paper ends with a summary of conclusions and references.

2   Semantic Web, Recommender Systems and Opinion Mining

Post-Via platform is based on the use of several technological elements. The most important elements from a technological and research perspective are the use of Semantic Technologies as key element for the representation of the knowledge and the inference of new knowledge, the recommender systems as artificial intelligence element which allows to computer the knowledge of the domain and the preference of an user to perform a personalized recommendation and the use of opinion mining techniques and tools to compute the reputation of the different touristic elements which are part of Post-Via platform.

In the following subsections a brief analysis of the state of the art of all this elements will be done.

2.1   Semantic Web

The Semantic Web is important area at the confluence of artificial intelligence (AI) and Web technologies; it proposes to add explicit description about the semantic of the resources, to allow the machines have a level of understanding of the web enough to do the most difficult part, which currently execute manually by user who interact with the web [13]. Nevertheless, the main obstacle is that, at present, the meaning of Web content is not machine-accessible. The Semantic Web extends the current Web providing machine processable semantics to Web resources [20]. However, some authors consider that the so-called deep web, the web pages that are dynamically generated from information stored in underlying knowledge bases, is even greater than the total volume of printed information existing in the world [52] and all this information is unstructured, therefore software agents (softbots) cannot understand and process this information, and much of the potential of the Web has so far remained untapped [18].

The Semantic Web relies heavily on formal ontologies to structure data for comprehensive and transportable machine understanding. Thus, the proliferation of ontologies factors largely in the Semantic Web's success [40] contributing to the generation of ontologies in several fields such as medical diagnosis [24][51], psychology [12], and enterprises [58] among several others. For a correct definition of the data, the Semantic web essentially uses RDF, SPARQL, and OWL technologies that help turn the Web into a global infrastructure where it is possible to share, and re-use data and documents between several types of users (W3C, 2010). A vast amount of information such as the existing in the current web which lacks of semantic annotation presents a problem several domains such as information search, information extraction, maintenance and automatic document generation [20]. For this reason, provide semantic annotation to the web using the aforementioned tools will allow solving these problems.

With the help of ontology, tourists' needs and preferences could be better understood, and the appropriate information and services resources could be located from the Semantic Web [14]. For this reason, and given the importance of the tourist sector, there are several works that are focused on e-tourism in the current literature. One interesting example is the system able to generate tourist packages in a dynamic way with the aim of create a personalized package depending on the preferences of the user, searching for different products and services, hotels, car rental, museums, theaters, etc. [11]. Also, Damljanović and Devedžić [16] have "developed Travel Guides - a prototype system for tourism management to illustrate how semantic web technologies combined with traditional E-Tourism applications: a.) help integration of tourism sources dispersed on the Web b) enable creating sophisticated user profiles. Maintaining quality user profiles enables system personalization and adaptivity of the content shown to the user. The core of this system is in ontologies – they enable machine readable and machine understandable representation of the data and more importantly reasoning."[16]. Another interested approach in the use of semantic technologies in tourist domain is the once provided by Ten-Hagen et al. [54], where the creation of DTG (Dynamic Tour Guide), a mobile agent which computes an itinerary of a couple of hours to explore a city. For this the DTG interrogates Tour Building Blocks (TBB), e.g. potential sights or restaurants, to determine current information, e.g. opening hours or availability. An ontology is used to capture the profiles of the TBBs and the interests of a tourist. Both are used by a semantic match algorithm to rank the TBBs.

Other examples on the use of semantic technologies in the tourism sector include Sem-Fit [25], a semantic based expert system to provide recommendations in the tourism domain, SPETA [23] and its later evolution to SPETA II [26], the LA_DMS project [35] provides semantic-based information for tourism destinations by combining the P2P paradigm with semantic web technologies or e-Tourism Planner-AuSTO, an ontology-based touristic planner that enables users to create an itinerary in one single application.

2.2   Recommender Systems

There are several technologies and systems which have emerged after the appearance of Internet. One of the main contributions of Internet is the appearance of the Web 2.0, which contributes to the collaboration among users. This collaboration cause an enormous generation of data over Internet. A study of the International Data Corporation (IDC) has estimated that in 2015, digital universe will take up about **8 Zettabytes** $\cong 8 \cdot 10^{12}$ **GB**, which states the remarkable expansion of Internet and the existing problem about information overload. For hence, due to the vast amount of information existing, it is necessary to develop techniques able to filter the available data, allowing us to consult only the information which is valuable for us [45]. Thus, the users of the online systems have the necessity of tools which helps them to face all the information available in the World-Wide-Web [2]. Recommender systems have demonstrated to be an answer to the problem of information overload, providing to the users more dynamic and personalized search services [44].

One of the fields where the recommender systems have been more applied is electronic commerce. So much so that there are authors which define recommender systems as a part in this new way of trade in Internet: "The recommender systems are intelligent e-commerce applications which helps the users in their searching tasks offering personalized recommendations during their interaction with the system" [1]. The inclusion of this environment is logic, because the recommendation of products or services to one user is very common. However, it is possible to find recommendation systems in other fields which don't aim an economical benefit such as news recommendation depending on the user profile [48] or users which we can know or follow on a social network [28].

At the beginning of ninety's first works about recommender systems could be found. These appear under the number of filtering systems with the aim of classify and select that news which can be interesting for a concrete user [22][50](Goldberg et al., 1992). Nowadays, there are several the works where we can found information about the different types of recommendation systems which exists such as social filter systems (or collaborative systems), recommendation systems based on content, or hybrid systems [1][8][41].

2.3   Opinion Mining

Two types of textual information live together nowadays in Internet: facts and opinions. First one are treated as events which can be tagged with keywords which allows search engines in the indexation process. However, the opinions are more complex.

The appearance of 2.0 platforms where it is allows making opinions about services and products has supposed a big change in the form of analyze the opinion of the users, making influence in the decision of other users. Some numbers which reflects this situation is that about 83% of the clients which buys tourism packages have been influenced by the comments of other customers (ChannelAdvisor "Consumer Shopping Habits Survey", August 2010). As is pointed out by Pang & Lee [46] the area of emotion analysis and opinion mining have experimented an enormous expansion in research activity. However, the interest on this topic is not new as is demonstrated by early investigations done by [9] or Wilks & Bien [66]. Later researches have been mostly focused in the interpretation of metaphors, narrative, points of view, affection and other areas [29][32][36][61][63][64][65].

In 2011 seems that a new trend emerged with the divulgation of the problems and advantages that opinion mining offers [10][17][39][42][43][47][55][56][62][67](Dave et al., 2003; Tateishi et al., 2001).

It is important to remark that an opinion is just a feeling, vision, attitude or evaluation, positive or negative about an entity or an aspect of an entity [31]. In this context, the formalization of an opinion can be divided in usual opinions and comparative opinions. Comparative opinions are the sentences or texts which compares more than one object or entity [33][34]. In this context, in opinion mining there are several important concepts which should be correctly defined in order to make a correct analysis of the emotions. For this reason it is important being able to differentiate between subjectivity, emotions and feelings, which are near but different concepts.

Opinion mining can be applied in several domains such as buying, entertainment, administration, research and development, education and marketing [4].

The aforementioned technologies play an important role in the development of Post-Via platform as have been mentioned before. The previous paragraphs reflects a brief perspective about each technology involved in Post-Via with the aim of allow a better understanding of the role which each technology plays in the context of Post-Via platform.

3   POST-VIA

POST-VIA 2.0 is a platform is based in the conviction that the application of information technologies (IT) can be applied to the tourist domain with the aim of improve the quality of the services offered by the providers to the final users. The aim of the platform is assist to the user before travel, during travel and after travel.

In this aim, POST-VIA 2.0 aims to reach several goals such as the management and analysis of the tourist travel experience, being able to extract useful knowledge which will be offered to the institutions responsible of the tourist policies (Destination Management Organizations (DMO), mainly) and to the individual providers of services, allowing to elaborate new marketing strategies based on the feedback of the tourists.

For these reasons POST-VIA 2.0 platform have been designed as a virtual meeting point for the different roles which are implied in the processes of management and provide touristic services. This platform will allow making relationships between those tourists which already finish their travel to a concrete destiny, and those tourists which are planning a new trip to the same destiny. At the same time, POST-VIA 2.0 will allow to the DMOs to have a direct relationships between both type of users (those who have finished their trip, and those who are planning a new one).

The extraction of the knowledge implies that the platform should include social components from Web 2.0 as well as other innovate elements for the realization of the information analysis, such as opinion mining, which will provide to the DMOS information about the tourist experience, allowing that the provided CRM 2.0 tools will become a trusty component for the elaboration of individualized marketing strategies.

The development of POST-VIA platform architecture is based on model-view-controller (MVC). The main advantage of this approach is that the development can be done in several levels, and for hence, if some change is needed, only the required level will be modified without the necessity of review and debug the code. This architecture is divided in three main components (MVC). **Fig. 1** presents the architecture of the current POST-VIA 2.0 platform.

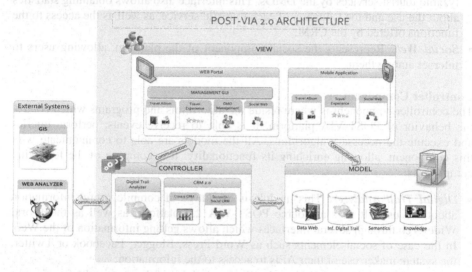

Fig. 1. Architecture of the current POST-VIA 2.0 platform

As can be seen the architecture of POST-VIA is based on the division of the platform in several modules which are inside the different parts of the MVC model. The communication of the POST-VIA 2.0 platform is done between the different modules allowing a rapid interchange of information. However, there are several external systems which also are used as part of the behavior of the POST-VIA 2.0 systems. The communication between the platform and these systems are generally done by the use of Web Services. In the next paragraphs, the components of POST-VIA will brief explained.

View Component:
The view component presents the system to the user through the graphical user interface (GUI). It allows the communication and capture of information from a graphical perspective. In the case of POST-VIA, the visualization of the system can be done from two different approaches: 1) from a traditional web platform or 2) from a mobile application.

- *GUI Management:* It is the interface which allows administering and managing the content of the platform depending on the role which is logged (DMOs or Users). The management GUI has been designed to take advantage of new web standards such as HTML5, CSS3, JQuery or AJAX, allowing the creation of Rich Internet Applications (RIAs). This interface only can be accessed from Web Portal.
- *Travel Album:* It is the interface of the application which allows creating a multimedia album associated to a travel or trip of the logged user.
- *Travel Experience:* This interface allows the valuation of the travel experience by the users as well as before the travel, during the travel or after the travel. It also offers enriched information such as recommendations.
- *DMO management:* It represents the interface which allows the update of the analyzable tourist services by the DMOs. This interface also allows obtaining statistics about the use and the valuation of each touristic service, as well as the access to the functions offered by the CRM.
- *Social Web:* Represents the social component of the platform, allowing users to interact among them.

Controller Component:
The controller component is where reside the algorithms and programs which control the behavior of POST-VIA platform. It is able to receive events, perform queries and execute the necessary algorithms. All the models are able to communicate with this component, allowing enriching its functionality. It is compose of the following components:

- *Digital Trail Analyzer:* This module is in charge of compile useful information about the user. It uses as source POST-VIA 2.0 platform as well as the World Wide Web through several interfaces which allows getting information of the Web. In the case of social elements such as Word Press, Blogger, Facebook or Twitter, the system makes use of their APIs to access to the information.
- *CRM 2.0:* This is one of the most important modules of the system. Its functionality is divided in two components.
  - o *Classic CRM:* It offers the basic services of a CRM component which will support the interests of the company with the necessary tools to manage the information of the platform.
  - o *Semantic CRM:* It is the most innovative component of the project and will be in charge of offer new marketing models to the DMOs. Several artificial intelligence techniques based on heuristics provided by the expert in the field of tourism will be used. In this component we can distinguish two necessary subcomponents:
    - ▪ *Opinion mining manager:* Using as data source the information provided by the users in their opinions it is possible, applying Natural Language Processing (NLP) techniques generate knowledge about several concepts associated to the domain.
    - ▪ *Inference engine:* The inference engine is in charge of providing an intelligent mechanism to infer new knowledge from the opinion entities extracted by opinion mining manager.

Model:
The model is where the data of the platform is stored. It is based on a database management system which is in charge of storing, access, modify and delete all the data of the platform. It receives queries from the controller or the view. Model also can throw requests to the controller to process the data and store it again after the processing task. It is composed by the following elements:

- **Data Web:** This is the element which more data diversity contains. It contains data and multimedia content managed by the DMOs. This data makes reference to the diverse destiny and tourist activities of a concrete location or area. It also contains information provided by the users, collected by the system in the travel experience module.
- **Digital Trail Information:** In this element all the data pertaining to the searching of digital trail will be stored, allowing other modules or algorithms to access to the information.
- **Semantic annotation:** Based on ontologies and knowledge representation techniques provides meaning to the concepts of the platform. It also stores relations about the different concepts which make up the platform.
- **Knowledge base:** Thanks to the information of the previous components a final knowledge base which embraces all the knowledge of the rest of modules can be generated, providing of a full-related knowledge module.

The presented architecture and their modules provide a set of functional behavior elements such as:

- Establishment of mechanisms which allow to the DMOs maintains the communication and the contact with the tourist once the visit has finished.
- Capture the perception, valuation and expectative of the tourists about the equipment, services and other aspects which make up the touristic offer.
- Establish attractive products and services for the tourists, inducing them to collaborate with the platform.

Based on the aforementioned objectives of the platform, POST-VIA will be work around the following fundamental components:

- Travel experience management (perception-valuation-analysis) through mechanisms which allows to know the perception that the tourists have about the different Point of Interest (POI), activities and touristic services which have experience during their travel.
- Activation of the social networks capabilities through the establishment of a personalized social network.
- Personal digital trail management through the marks that a tourist leaves when is making different activities on a controlled physical environment with the corresponding digital representation.
- Elaboration of CRM and social semantic CRM tools which makes possible new marketing models, creation tools for the DMOs to consult and manage all the information derived of the application of the previous lines, allowing elaborating personalized marketing strategies depending on the tourist experience.

- Integration with relevant external systems which allows providing to the users new interesting functionalities: location and spatial analysis from Geographic Information Systems (GIS) or Location Based Services (LBS) or connection with external recommender systems.

To summarize, POST-VIA platform can be seen a new challenge in the application of Information Technology (IT) to the tourist domain, allowing to improve the experience of travelling having a complete support during all the travel process (before, during and after). The development of the web platform allows the user to manage all the information associated to the travel, allowing experimenting a new perspective. The mobile application which have been developed to be used with the main mobile operating systems allows to the user not only to make a rating about the services offered by a tourist provider or point of interest; the user also can have real-time information about the offers promoted by the DMOs or the POIs manager.

Finally, the interaction of the users with the platform represents an enormous advantage to the providers of tourist services because they can access to the information stored by the tourist through the use of the platform, allowing the providers to prepare new marketing strategies based on the preference of the tourists and also to prepare personalized advertisements and offers depending on the tourist preferences.

4   Conclusions and Future Work

The Internet has disrupted traditional tourism in which a promising landscape of intelligent service provision has erupted by applying a new lattice of cutting-edge technologies [26]. This new environment leads to a new way to manage the provision of tourism services and to monitor services in a changing scenario. Now tourism stakeholders are forced to operate in a coordinated manner to increase the value of tourism, to keep current tourists and attract new ones. The Post-Via framework provides a new platform and architecture to attract tourist to places in which they have been before. Based on geographical information systems, recommender systems, customer relationship management, social networks, mobility and semantic technologies, Post-Via represents a novel approach to achieve tourism loyalty.

There are many tools devoted to provide tourist services heavily dependent on ICT. The models and tools make possible to obtain tourist information, to plan itineraries and to make reservations through web portals. However, there are far fewer models devoted to post-trip issues. This stage of the tourist experience is crucial for proper management of demand and, among other things, encourages loyalty tourists about the destination just visited. The actual knowledge of that travel experience becomes a privileged tool to enable public authorities responsible for tourism policy (DMOs) and individual service providers rework and refine business strategies based on the feedback provided by the tourists themselves.

Post-Via fills this gap. It captures and effectively manages all this knowledge, using a selective and optimized service delivery post-trip to the various tourism stakeholders involved. Basic differential element, the VIA POST 2.0 platform is not limited to rely on the goodwill of tourists (often controversial and always random) to fill the valuable data of subjective perception, but offers an attractive product catalog and services compelling enough to take the time and the interest to collaborate.

Acknowledgements. This work is supported by the Spanish Ministry of Industry, Tourism, and Commerce under the project "Diseño, desarrollo y prototipado de una plataforma TIC de servicios post-viaje a los turistas" (IPT-2011-0973-410000).

References

1. Adomavicius, G., Tuzhilin, A.: Toward the next generation of recommender systems: A survey of the state-of-the-art and possible extensions. IEEE Transactions on Knowledge and Data Engineering 17(6), 734–749 (2005)
2. Balabanovic, M., Shoham, Y.: Fab: content-based, collaborative recommendation. Communications of the ACM 40(3), 66–72 (1997)
3. Benjamins, V.R., Contreras, J., Corcho, O., Gómez-Pérez, A.: The six challenges of the Semantic Web. In: Eighth International Conference on Principles of Knowledge Representation and Reasoning (2002)
4. Binali, H., Potdar, V., Wu, C.: A state of the art opinion mining and its application domains. In: IEEE International Conference on Industrial Technology, ICIT 2009, pp. 1–6 (2009)
5. Buhalis, D., Law, R.: Progress in information technology and tourism management: 20 years on and 10 years after the Internet—The state of eTourism research. Tourism Management 29(4), 609–623 (2008)
6. Buhalis, D.: eTourism: Information technology for strategic tourism management. Pearson (Financial Times/Prentice-Hall) (2003)
7. Buhalis, D.: cAirlines: strategic and tactical use of ICTs in the airline industry. Information & Management 41(7), 805–825 (2004)
8. Candiller, L., Jack, K., Fessant, F., Meyer, F.: State-of-the-Art Recommender Systems. In: Collaborative and Social Information Retrieval and Access Techniques for Improved User Modeling, pp. 1–22 (2009)
9. Carbonell, J.: Subjective Understanding: Computer Models of Belief Systems. PhD thesis, Yale (1979)
10. Cardie, C., Wiebe, J., Wilson, T., Litman, D.: Combining low-level and summary representations of opinions for multi-perspective question answering. In: Proceedings of the AAAI Spring Symposium on New Directions in Question Answering, pp. 20–27 (2003)
11. Cardoso, J.: Developing Dynamic Packaging Systems using Semantic Web Technologies. Transactions on Information Science and Applications 3(4), 729–736 (2006)
12. Casado-Lumbreras, C., Rodríguez-González, A., Alvarez-Rodríguez, J.M., Colomo-Palacios, R.: PsyDis: Towards a diagnosis support system for psychological disorders. Expert Systems With Applications (2012),
http://dx.doi.org/10.1016/j.eswa.2012.04.033
13. Castells, P.: La Web Semántica. Sistemas Interactivos y Colaborativos en la Web, Ediciones de la Universidad de Castilla-La mancha, 195–212 (2003)
14. Chiu, D.K., Cheung, S.C., Leung, H.F.: A Multi-Agent Infrastructure for Mobile Workforce Management. Service Oriented Enterprise. In: Proc. HICSS38. IEEE Computer Society Press, Big Island (2005)

15. Chung, J.Y., Buhalis, D.: Virtual Travel Community: bridging between travellers and locals. In: Sharda, N. (ed.) Tourism Informatics: Visual Travel Recommender Systems, Social Communities and User Interface Design. Information Science Reference, pp. 130–144 (2009)
16. Damljanovic, D., Devedzic, V.: Applying semantic web to e-tourism. In: Ma, Z. (ed.) The Semantic Web for Knowledge and Data Management: Technologies and Practices. IGI Global (2008)
17. Das, S., Chen, M.: Yahoo! for Amazon: Extracting market sentiment from stock message boards. In: Proceedings of the Asia Pacific Finance Association Annual Conference, APFA (2001)
18. Doan, A., Madhavan, J., Dhamankar, R., Domingos, P., Halevy, A.: Learning to match ontologies on the Semantic Web. The VLDB Journal 12(4), 303–319 (2003)
19. Dwyer, L., Edwards, D., Mistilis, N., Roman, C., Scott, N.: Destination and enterprise management for a tourism future. Tourism Management 30(1), 63–74 (2009)
20. Fensel, D., Facca, F.M., Simperl, E., Toma, I.: Semantic Web. Semantic Web Services, 87–104 (2011)
21. Fensel, D., van Harmelen, F., Horrocks, I., McGuinness, D.L., Patel-Schneider, P.F.: OIL: an ontology infrastructure for the Semantic Web. IEEE Intelligent Systems 16(2), 38–45 (2001)
22. Foltz, P.W., Dumais, S.T.: Personalized information delivery: an analysis of information filtering methods. Communications of the ACM 35(12), 51–60 (1992)
23. García Crespo, A., Chamizo, J., Rivera, I., Mencke, M., Colomo Palacios, R., Gómez Berbís, J.M.: SPETA: Social pervasive e-Tourism advisor. Telematics and Informatics 26(3), 306–315 (2009)
24. García-Crespo, A., Colomo-Palacios, R., Gómez-Berbís, J.M., Chamizo, J., Rivera, I.: Intelligent Decision-Support Systems for e-Tourism: Using SPETA II as a Knowledge Management Platform for DMOs and e-Tourism Service Providers. International Journal of Decision Support System Technology 2(1), 35–47 (2010)
25. García-Crespo, Á., López-Cuadrado, J.L., Colomo-Palacios, R., González-Carrasco, I., Ruiz-Mezcua, B.: Sem-Fit: A semantic based expert system to provide recommendations in the tourism domain. Expert Systems with Applications 38(10), 13310–13319 (2011)
26. García-Crespo, A., Rodríguez, A., Mencke, M., Gómez-Berbís, J.M., Colomo-Palacios, R.: ODDIN: Ontology-driven differential diagnosis based on logical inference and probabilistic refinements. Expert Systems with Applications 37(3), 2621–2628 (2010)
27. Gianforte, G.: The World at Our Fingertips - How Online Travel Companies Can Turn Clicks into Bookings. Journal of Vacation Marketing 10(1), 79–86 (2003)
28. Hannon, J., Bennett, M., Smyth, B.: Recommending twitter users to follow using content and collaborative filtering approaches. In: Proceedings of the Fourth ACM Conference on Recommender Systems (RecSys 2010), pp. 199–206. ACM, New York (2010)
29. Hearst, M.: Direction-based text interpretation as an information access refinement. In: Jacobs, P. (ed.) Text-Based Intelligent Systems, pp. 257–274. Lawrence Erlbaum Associates (1992)
30. Hjalager, A.M.: A review of innovation research in tourism. Tourism Management 31(1), 1–12 (2010)
31. Hu, M., Liu, B.: Mining and Summarizing Customer Reviews. In: KDD, Seattle (2004)
32. Huettner, A., Subasic, P.: Fuzzy typing for document management. In: ACL 2000 Companion Volume: Tutorial Abstracts and Demonstration Notes, pp. 26–27 (2000)

33. Jindal, N., Liu, B.: Identifying Comparative Sentences in Text Documents. In: Proceedings of the 29th Annual International ACM SIGIR Conference on Research and Development in Information Retrieval (2006a)
34. Jindal, N., Liu, B.: Mining Comparative Sentences and Relations. Mining Comparative Sentences and Relations (2006b)
35. Kanellopoulos, D.N.: An ontology-based system for intelligent matching of travellers' needs for Group Package Tours. International Journal of Digital Culture and Electronic Tourism 1(1), 76–99 (2008)
36. Kantrowitz, M.: Method and apparatus for analyzing affect and emotion in text. U.S. Patent 6622140 (2003)
37. Kenteris, M., Gavalas, D., Economou, D.: An innovative mobile electronic tourist guide application. Personal and Ubiquitous Computing 13(2), 103–118 (2009)
38. Kusha, D., Lawrence, S., Pennock, D.M.: Mining the peanut gallery: Opinion extraction and semantic classification of product reviews. In: Proceedings of WWW, pp. 519–528 (2003)
39. Liu, H., Lieberman, H., Selker, T.: A model of textual affect sensing using real-world knowledge. In: Proceedings of Intelligent User Interfaces (IUI), pp. 125–132 (2003)
40. Maedche, A., Staab, S.: Ontology learning for the Semantic Web. IEEE Intelligent Systems 16(2) (2001)
41. Mooney, R.J., Roy, L.: Content-based book recommending using learning for text categorization. In: Proceedings of the fifth ACM conference on Digital libraries, DL 2000 (2000)
42. Morinaga, S., Yamanishi, K., Tateishi, K., Fukushima, T.: Mining product reputations on the web. In: Proceedings of the ACM SIGKDD Conference on Knowledge Discovery and Data Mining (KDD), pp. 341–349 (2002)
43. Nasukawa, T., Yi, J.: Sentiment analysis: Capturing favorability using natural language processing. In: Proceedings of the Conference on Knowledge Capture, K-CAP (2003)
44. O'Donovan, J., Smyth, B.: Trust in recommender systems. In: Proceedings of the 10th international conference on Intelligent user interfaces (IUI 2005), pp. 167–174 (2005)
45. Sullivan, D.O., Wilson, D.C., Smyth, B.: Improving case-based recommendation: A collaborative filtering approach. In: Craw, S., Preece, A.D. (eds.) ECCBR 2002. LNCS (LNAI), vol. 2416, pp. 278–291. Springer, Heidelberg (2002)
46. Pang, B., Lee, L.: Opinion Mining and sentiment analysis. Foundations and Trends in Information Retrieval 2(1-2), 1–135 (2008)
47. Pang, B., Lee, L., Vaithyanathan, S.: Thumbs up? Sentiment classification using machine learning techniques. In: Proceedings of the Conference on Empirical Methods in Natural Language Processing (EMNLP), pp. 79–86 (2002)
48. Phelan, O., McCarthy, K., Smyth, B.: Using twitter to recommend real-time topical news. In: Proceedings of the Third ACM Conference on Recommender Systems (RecSys 2009), pp. 385–388 (2009)
49. PhoCusWright, U.S. Online Travel Overview, 11th edn.(2011)
50. Resnick, P., Iacovou, N., Suchak, M., Bergstrom, P., Riedl, J.: GroupLens: an open architecture for collaborative filtering of netnews. In: Proceedings of the 1994 ACM Conference on Computer Supported Cooperative Work (CSCW 1994), pp. 175–186 (1994)
51. Rodríguez-González, A., Hernandez-Chan, G., Colomo-Palacios, R., Gomez-Berbís, J.M., García-Crespo, A., Alor-Hernandez, G., Valencia-Garcia, R.: Towards an Ontology to support semantics enabled Diagnostic Decision Support Systems. Current Bioinformatics (2012a) (in press)

52. Ruiz-Casado, M., Alfonseca, E., Castells, P.: Automatising the learning of lexical patterns: An application to the enrichment of WordNet by extracting semantic relationships from Wikipedia. Data & Knowledge Engineering 61(3), 484–499 (2007)
53. Stockdale, R.: Managing customer relationships in the self-service environment of e-tourism. Journal of Vacation Marketing 13(3), 205–219 (2007)
54. Ten-Hagen, K., Kramer, R., Hermkes, M., Schumann, B., Mueller, P.: Semantic Matching and Heuristic Search for a Dynamic Tour Guide. Information and Communication Technologis in Tourism 5, 149–159 (2005)
55. Tong, R.M.: An operational system for detecting and tracking opinions in on-line discussion. In: Proceedings of the Workshop on Operational Text Classification, OTC (2001)
56. Turney, P.: Thumbs up or thumbs down? Semantic orientation applied to unsupervised classification of reviews. In: Proceedings of the Association for Computational Linguistics (ACL), pp. 417–424 (2002)
57. UNWTO. UNWTO, eBusiness for tourism: Practical guidelines for destinations and businesses, World Tourism Organisation, Madrid (2001)
58. Uschold, M., King, M., Moralee, S., Zorgios, Y.: The enterprise ontology. The Knowledge Engineering Review 13(1) (1998)
59. Vogt, C.A.: Customer Relationship Management in Tourism: Management Needs and Research Applications. Journal of Travel Research 50(4), 356–364 (2011)
60. Werthner, H., Ricci, F.: E-commerce and tourism. Commun. ACM 47(12), 101–105 (2004)
61. Wiebe, J., Bruce, R.: Probabilistic classifiers for tracking point of view. In: Proceedings of the AAAI Spring Symposium on Empirical Methods in Discourse Interpretation and Generation, pp. 181–187 (1995)
62. Wiebe, J., Breck, E., Buckley, C., Cardie, C., et al.: Recognizing and organizing opinions expressed in the world press. In: Proceedings of the AAAI Spring Symposium on New Directions in Question Answering (2003)
63. Wiebe, J.M.: Identifying subjective characters in narrative. In: Proceedings of the International Conference on Computational Linguistics (COLING), pp. 401–408 (1990)
64. Wiebe, J.M.: Tracking point of view in narrative. Computational Linguistics 20(2), 233–287 (1994)
65. Wiebe, J.M., Rapaport, W.J.: A computational theory of perspective and reference in narrative. In: Proceedings of the Association for Computational Linguistics (ACL), pp. 131–138 (1988)
66. Wilks, Y., Bien, J.: Beliefs, points of view and multiple environments. In: Proceedings of the International NATO Symposium on Artificial and Human Intelligence, pp. 147–171 (1984)
67. Yu, H., Hatzivassiloglou, V.: Towards answering opinion questions: Separating facts from opinions and identifying the polarity of opinion sentences. In: Proceedings of the Conference on Empirical Methods in Natural Language Processing, EMNLP (2003)

A Help Desk Support System Based on Relationship between Inquiries and Responses

Masaki Samejima[1], Masanori Akiyoshi[2], and Hironori Oka[3]

[1] Osaka University, 2-1 Yamadaoka, Suita-shi, Osaka, 565-0871, Japan
samejima@ist.osaka-u.ac.jp
[2] Hiroshima Institute of Technology,
2-1-1 Miyake, Saeki-ku, Hiroshima-shi, Hiroshima, 731-5193, Japan
m.akiyoshi.we@cc.it-hiroshima.ac.jp
[3] Codetoys K.K., 2-6-8 Nishitenma, Kita-ku, Osaka-shi, Osaka, 530-0047, Japan
oka@codetoys.co.jp

Abstract. We propose a help desk support system to extract FAQ (Frequently Asked Questions) automatically, to retrieve FAQ for inquiry e-mails, and to show FAQ in users' inputting their inquiry e-mail. In the help desk, operators record inquiry e-mails and response e-mails. Between inquiries and response, there are relationship that similar words are often used. First, we propose a classification method of inquiry e-mails for describing FAQ with pairs of inquiries and responses. Second, we propose a detection method of the FAQ matching inquiry e-mail based on Jaccard coefficient between inquiries and responses. Finally, we propose a predictive search method of FAQ by matching an incomplete inquiry to FAQ.

Keywords: Help desk, FAQ, Clustering, Jaccard coefficient, Retrieval.

1 Introduction

Recently web-based services, for instance, shopping and community management, are rapidly increasing, which usually provide basic interactions between users and companies by the form-based input on web pages. Inquiry e-mails from users are delivered to company's operators and they send back e-mails. Along with lots of inquiries, FAQ pages are introduced to reduce such operator tasks [16,21,6,5,12].

FAQ pages, however, has not sufficiently reduced the operator tasks, which is caused by two factors. One factor is that users send e-mails without checking FAQ pages. There are great deal of inquiries related to FAQ on web pages and operators must process such inquiries. Another factor is the lack of inquiry and response sets on current FAQ. To make a new inquiry and response set, operators must analyze inquiry records, extract frequent questions, and construct a proper inquiry and response set. It takes great deal of time for operators to look through a large number of inquiries.

We address finding FAQ to the inquiry e-mails and building up FAQ automatically. There are relationships between inquiry and response e-mails. For

T. Matsuo & R. Colomo-Palacios (Eds.): *Electronic Business and Marketing*, SCI 484, pp. 43–66.
DOI: 10.1007/978-3-642-37932-1_5   © Springer-Verlag Berlin Heidelberg 2013

example, some of inquiries and corresponding responses are similar to each other. So, we propose a help desk support system based on relationship between inquiries and responses. The proposed system consists of a classification method of inquiry e-mails for describing FAQ, a detection method of FAQ matching inquiry e-mails, and a predictive search method of FAQ by matching an incomplete inquiry to FAQ.

2   Outline of the Online Help Desk

2.1   Problems of the Online Help Desk

The purpose of the help desk is to answer the inquiries from users. Recently, the function of the help desk is provided on the web-based system. First, we discuss the model of the help desk shown in Fig. 1 [13,3].

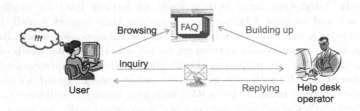

Fig. 1. Process model of the help desk

When users have some questions, they browse FAQ or send inquiry e-mails to the operators in the help desk. The tasks of the online help desk is to reply to the inquiries and to build up FAQ:

– Replying to inquiries
  Response to inquiries from users is the most fundamental task of the help desk. However it takes great deal of time for operators to process users e-mails. Because of the variety of users demands and company services, the increasing number of inquiries has been processed by operators.
– Building up FAQ
  Building up FAQ is also an important task of the help desk. Companies build up and post FAQ on their web sites to reduce the number of inquiries from users. Help desk managers analyze records of the inquiry and response e-mails, find appropriate sets of questions and answers, and adds the sets to their FAQ pages [4].

2.2 Research Purpose

The reason why many inquiries are sent to the help desk is that users do not check FAQ before they send inquiries. According to our preliminary investigation, 30 to 40 percents of inquiries are related to FAQ. Operators can make responses to the inquiries easily by using FAQ. So, one of the research purposes is to find automatically FAQ for each inquiry.

Additionally, in order to reduce the effort involved in building up FAQ, another research purpose is to build up FAQ based on past inquiry and response e-mails.

3 Help Desk Support System Based on Relationship between Inquiries and Responses

Because the help desk has FAQ for replying to the inquiries efficiently, the proposed method also uses FAQ to support the help desk. However, inquiries from users are not always related to FAQ that is summarized from the past inquiries and responses. By using the past inquiries and responses, the proposed method realize the display of FAQ to users [10]. The response is made by an operator based on an inquiry. So the inquiry and the response are often related to each other. We analyze the relationship between the inquiry and the response and propose a help desk support system based on the relationship. The outline of our help desk support system is shown in Fig. 2.

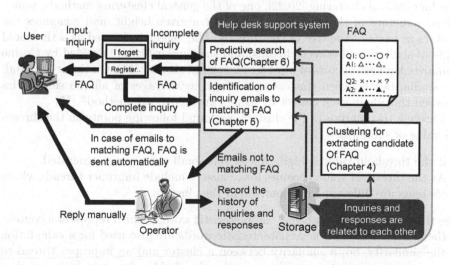

Fig. 2. Outline of the help desk support system

In order to prevent users from sending inquiries that match to FAQ, the proposed system displays FAQ when a user is inputting the inquiry to the web form. By providing a predictive search function based on the user's incomplete inquiry, the proposed system can show FAQ that probably corresponds to the inquiry, which allow users to know the solution in inputting the inquiry.

After the user inputs the inquiry completely and sends it to the operator, the correct response is guaranteed for the inquiry . So, the proposed system identifies inquiry e-mails to exactly matching FAQ in order to reply automatically without operators. If the proposed system judges that the inquiry does not match any FAQ, the inquiry is sent to the operators.

When the operators reply the inquiry that does not match any FAQ, the inquiries and the responses are recorded in the storage of the proposed system. For making FAQ from the responses and the inquiries in the storage, the proposed method makes clusters of the pairs of an inquiry and a response as candidates of FAQ. Here, we call the pair "thread". Checking only the candidates of FAQ, the operators can maintain FAQ.

In this paper, a stepwise clustering method for extracting candidate FAQ is described in chapter 4, an identification method of inquiry e-mails to the matching FAQ is described in chapter 5, and a predictive search method of FAQ by statistical model of important words occurrence is described in chapter 6.

4   A Stepwise Clustering Method for Extracting Candidate FAQ

4.1   Outline of Stepwise Clustering

The hierarchical clustering [20,22], one of the general clustering methods, builds a tree structure of threads, cuts the tree at a given height, and generates the clusters as parts of the tree of the threads. The height is decided as a threshold value of similarities between threads, and the similarities are decided by Cosine similarity between vectors of word frequencies in threads [17]. On the other hand, the similarities between clusters are defined as averages of all the similarities between threads in each cluster by using "group average method" [15] .

Through the analysis of the clusters, we found following points on the threshold value of similarities [18]:

– If the threshold value is high, precise but small clusters are generated.
– As the threshold value becomes low, clusters include improper threads whose contents are different from contents of the clusters.

The threads in a cluster include "characteristic words" which represent a content of the cluster. However, non-characteristic words are also used for a calculation of the similarity. So, a similarity between a cluster and an improper thread to a content of the cluster may be over the threshold value, which causes that the cluster can contain the improper thread. Therefore, we propose a clustering method by reflecting characteristics of words to the similarity. The proposed method uses "category dictionary" that has values indicating how characteristic the words in each cluster are [7,9]. In the dictionary, characteristic words have high values, and non-characteristic ones have small values. These are weighted

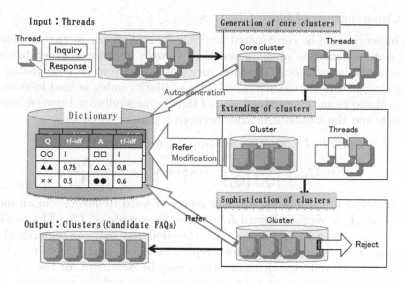

Fig. 3. Overview of clustering with dictionary

to the similarity so as to reflect the characteristics. In order to generate precise clusters, the dictionary needs to have enough words and appropriate values of weights for the words. However, the construction of the dictionary is time-consuming task for operators. So, it is necessary to generate clusters and update the dictionary automatically and accurately.

Fig. 3 shows the flow of extracting candidates of FAQ by clustering method that consists of the following three steps:

(1) Making core clusters by a high strictly threshold value: In order to ensure the accuracy at the beginning of the clustering, the small but precise clusters (core clusters) are generated by hierarchical clustering with a high threshold value. And values in the dictionary are decided as *tf-idf*(term frequency inverse document frequency) : words' typical indicators for characteristics [14].

(2) Expanding clusters by an appropriately-loosened low threshold value: The small cluster is not regarded as candidate FAQ, because it is thought that the content of the small cluster is not a frequent inquiry. Therefore, core clusters are expanded with a low threshold value by referring the category dictionary.

(3) Cleansing clusters: Improper threads in a cluster are removed from the cluster.

Theses three steps need thresholdvalues, which are impracticable to set appropriately by hand. Therefore we also propose an automatic setting mechanism of these threshold values.

4.2   Construction of Core Clusters

Core clusters should be constructed precisely for making the dictionary that has appropriate information of characteristics of words in order to generate correct clusters in the later steps. Therefore, core clusters have to be constructed with strictly similar threads to each other. This similarity index is used in clustering and calculated from the weighted sum of the Cosine similarity between inquiries of threads and the Cosine similarity between replies of threads.

$$Sim(Th_i,Th_j)=(1-\alpha)cosSimQ_{i,j}+\alpha cosSimA_{i,j} \tag{1}$$

$$cosSimQ_{i,j}=\frac{Q_i \cdot Q_j}{\|Q_i\|\,\|Q_j\|}, \quad cosSimA_{i,j}=\frac{A_i \cdot A_j}{\|A_i\|\,\|A_j\|}$$

Th_i is a thread of Q_i and A_i, Q_i is a vector of word frequencies in an inquiry of Th_i, and A_i a vector of word frequencies in a reply of Th_i. The similarity index is derived as $Sim()$, $cosSimQ_{i,j}$ is the similarity between inquiries Q_j, Q_i, $cosSimA_{i,j}$ is the similarity between replies of A_j, A_i and $\alpha(0 < \alpha < 1)$ is a constant value to reflect which similarities may be dominant for the clustering. The replies are usually written by specific operators and the words used in the replies of the same content are similar. Therefore α might be larger than 0.5.

After the construction of core clusters, a category dictionary is generated from the core clusters. This category dictionary is referred in the expansion and sophistication of clusters. The category dictionary keeps *tf-idf* value of each word in each cluster as a typical indicator for characteristics of each cluster. A *tf-idf* value of $Word_s$ gets a high value if the word appears frequently in the thread Th_i and the number of clusters containing the word is small.

$$tf\text{-}idf(Th_i, Word_s) = tf_{i,s} \times idf_s$$

$$tf_{i,s} = \frac{\text{Freq. of } Word_s \text{ in } Th_i}{\text{Num. of all words in } Th_i}$$

$$idf_s = \log\frac{\text{Num. of all clusters}}{\text{Num. of clusters including } Word_s}$$

A help desk operator decides the threshold value in this step so as to satisfy that core clusters contain strictly similar threads to each other and contents of core clusters are exclusive. The average of similarities between threads in the cluster is useful for judging that the cluster consists of similar threads. If the average is high, the operator can grasp that a core cluster contains strictly similar threads.

In order to judge that the contents of the core clusters are exclusive, we defined the similarity subtracted from 1.0 as an orthogonal index between core clusters. Fig. 4 shows a change of the average of the orthogonal indices with a change of the threshold value. Differential values of the averages gradually converge on 0, as the value of the threshold value is decreased. An operator can set the threshold value with the convergence because the convergence means that contents of core clusters get exclusive.

From these two points, an operator sets the threshold value generating core clusters which contain strictly similar threads and whose contents are exclusive to other clusters' contents.

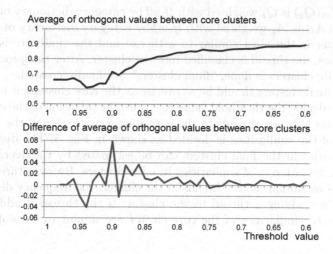

Average of orthogonal values between core clusters

Difference of average of orthogonal values between core clusters

Threshold value

Fig. 4. Change of orthogonal indices

4.3 Expansion of Clusters

In this step, core clusters constructed in the first step are expanded for extracting candidates of FAQ. The process of the cluster expansion with the dictionary is executed by adding a thread to a cluster and by combining two clusters.

Because core clusters are constructed with strictly similar threads to each other, a lot of threads are not included in any clusters. These threads outside core clusters should be added to a similar cluster based on the category dictionary. Furthermore it is necessary to combine similar clusters. The dictionary is updated at every expansion so as to put current information in it.

Adding a Thread to a Cluster. Because threads in a cluster of candidate FAQ must be similar to each other, a similar thread to the cluster can be added to the cluster. A similarity between a cluster and a thread is decided by the following formula:

$$Sim(Cluster_m, Th_j) =$$
$$(1 - \alpha)\, cosSimQ_{m,j} + \alpha\, cosSimA_{m,j} \qquad (2)$$
$$cosSimQ_{m,j} = \sum\nolimits_{i=1}^{n} cosSimQ_{i,j}\,/n$$
$$cosSimA_{m,j} = \sum\nolimits_{i=1}^{n} cosSimA_{i,j}\,/n$$
$$cosSimQ_{i,j} = \frac{tf\text{-}idf_m(Q_i) \cdot tf\text{-}idf_m(Q_j)}{\|tf\text{-}idf_m(Q_i)\|\,\|tf\text{-}idf_m(Q_j)\|}$$
$$cosSimA_{i,j} = \frac{tf\text{-}idf_m(A_i) \cdot tf\text{-}idf_m(A_j)}{\|tf\text{-}idf_m(A_i)\|\,\|tf\text{-}idf_m(A_j)\|}$$

where $tf\text{-}idf_m(Q_i)$ is Q_i weighted with $tf\text{-}idf$ by category dictionary of $cluster_m$ and $tf\text{-}idf_m(A_i)$ is A_i weighted with $tf\text{-}idf$ by category dictionary of $cluster_m$. When a cluster is the most similar to a thread and the similarity is over the threshold value, the thread is classified into the cluster. After this process for all threads outside clusters is done, "final clusters" are finally created.

While the final clusters should be as precise as the core clusters, it is also necessary how to decide the threshold value to ensure the precision of the expansion. The similarities between threads in the core cluster are high and the frequency distribution of the similarities is decided as shown in Fig. 5. The distributions of the similarities in the final clusters can be estimated by the average μ and the standard deviation σ of similarities in core clusters. Because threads to be added are not as similar as the threads in the cluster, the frequency distribution is changed after adding a thread to the cluster. If the thread is added to the cluster correctly, similarities of the core cluster are similar to ones of the final cluster.

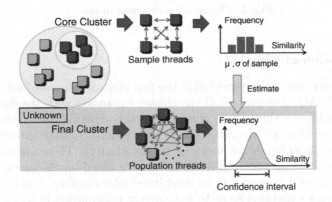

Fig. 5. Estimating population from core cluster

So when the frequency distribution of similarities changes after the expansion, the proposed method judges whether an added thread is correct or not by statistical testing that the clusters before and after the expansion can be regarded as the same.

For estimating the distributions of the final clusters, the average μ and the standard deviation σ are necessary. The proposed method derives μ and σ from the similarities in the $cluster_m$ by the following formula:

$$Sim_{Cluster_m}(Th_i, Th_j) =$$
$$(1 - \alpha)\, cosSimQ_{i,j} + \alpha\, cosSimA_{i,j} \qquad (3)$$

"Confidential interval" of the average of similarities are used as the threshold values of the clustering. Fig. 6 shows the judgment whether adding a thread is stopped or continued. The threshold values are decided to be the lower confidence limit. If the average of similarities in the expanded cluster is lower than the threshold value, adding a thread to the cluster is stopped. When adding to all clusters is stopped, this process is ended. Then the proposed method can set the threshold value individually and automatically for each cluster.

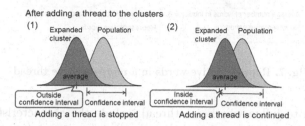

Fig. 6. Judgment whether adding a thread is stopped or continued

Combining Two Clusters. Because the threshold value in the step of constructing core clusters is a high value, a lot of small clusters can be constructed. Similar clusters have to be combined for acquiring large clusters as candidates of FAQ. A similarity between clusters is calculated by using *tf-idf* in the category dictionary. There are non-characteristic words that have small *tf-idf*, which makes similarities higher even if the contents are not similar. So, words in the top k of *tf-idf* are used for deciding similarities as the following formula:

$$Sim(Cluster_m, Cluster_n) = $$
$$\frac{tf\text{-}idf_{Q_m}[k] \cdot tf\text{-}idf_{Q_n}[k] + tf\text{-}idf_{A_m}[k] \cdot tf\text{-}idf_{A_n}[k]}{2} \tag{4}$$

where $tf\text{-}idf_{Q_m}[k]$ and $tf\text{-}idf_{A_m}[k]$ are vectors having upper k elements of inquiry and reply in category dictionary of $cluster_m$ respectively. And k is decided as follows. Firstly, the accumulated average of the top i of *tf-idf* in category dictionary of the $cluster_m$ is calculated. Because *tf-idf* of non-characteristic words are small and not so different, the accumulated average converges to 0 as i is increased. So, the proposed method calculates the second difference of the accumulated average and selects k when the second difference converges on 0.

If the highest similarity is more than the threshold value given in advance, the two clusters are combined.

4.4 Sophistication of Clusters

The pre-process may add threads to a improper cluster because *tf-idf* values of characteristic words in the dictionary are not completely calculated. After the construction of the dictionary is completed, the proposed method can judge whether an added thread is proper or improper to the cluster. So, the proposed method removes threads that do not include the characteristic words of the cluster.

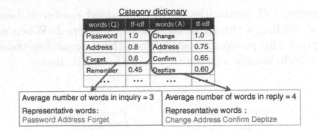

Fig. 7. Representative words in a representative thread

As a criteria to judge whether a thread include the characteristic words or not, the method generates a virtual thread called "representative thread" that includes just all characteristic words in the cluster. In order to generate a representative thread, the upper m words on *tf-idf* in the category dictionary are chosen as Fig. 7 shows. Then m is decided as an average number of words in threads in a cluster. The method decides whether threads in a cluster should be removed by Cosine similarity with the representative thread. If the similarity is lower than a threshold value, the thread is removed.

To set the threshold value in cleansing clusters individually and automatically, we use the average of similarity with the representative thread. The average of similarities with the representative thread has a relation to the similarities between the representative thread and the threads in a cluster. If the cluster is not precise, the similarities with the representative thread are low. So, the threshold value of $cluster_m$ is the value which is the standard deviation (σ_m) of similarity in each cluster subtracted from the average of similarity (μ_m):

$$Threshold_m = \mu_m - \sigma_m$$

5   An Identification Method of Inquiry E-mails to the Matching FAQ

5.1   Approach for Automatic Question Answering

As existing methods [11], Cosine similarity, Jaccard coefficient and SVM(Support Vector Machine)[2,19] are used for identifying similar documents. If Jaccard coefficient is applied to this problem, it may be possible to detect FAQ which is the most similar to the inquiry as FAQ which has the highest Jaccard coefficient.

Fig. 8 and Fig. 9 show Jaccard coefficient of some inquiry sentences and FAQs. The inquiries shown in solid lines in the graphs are the case of taking maximum Jaccard coefficient between inquiry sentence and correct FAQs. On the other hand, the inquiries shown in dashed lines are the cases of NOT taking maximum Jaccard coefficient between them. These graphs show that it is difficult to choose a correct FAQ simply by Jaccard coefficient. Cosine similarity has also the same

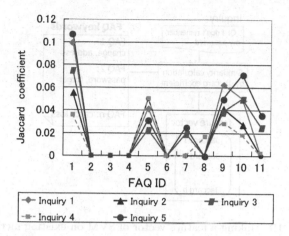

Fig. 8. Jaccard coefficients of the inquiry matching FAQ "1" to each FAQ

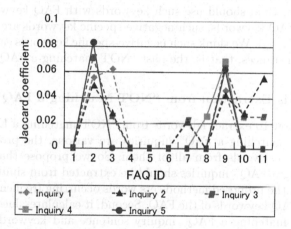

Fig. 9. Jaccard coefficients of the inquiry matching FAQ "2" to each FAQ

problem. Meanwhile there are some inquiries that match a FAQ by Jaccard coefficient, so that Jaccard coefficient can be the indicator of finding a matching FAQ. As shown in Fig. 10, the existing method using SVM sets keywords on each FAQ in advance and makes a feature vector whose elements are Jaccard coefficients with an inquiry and keywords of each FAQ. However, the feature vector does not reflect characteristics of negative instances. So this method did not get sufficient results on discriminating inquiries which are quite similar but "NOT matching" to a certain FAQ.

We propose a method to reflect characteristics of negative instances. Our approach comes from investigating the inquiries when help desk operators descriminate a set of inquiries that are quite similar to the contents of a certain FAQ but "NOT matching". "NOT matching a FAQ" inquiries have mostly some

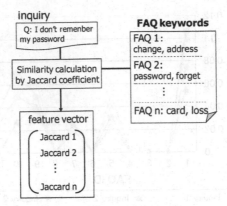

Fig. 10. Making a feature vector of SVM on existing method

specific words in addition to FAQ keywords, we call, negative-specific keywords. Therefore our method should use such keywords with FAQ keywords. However different from FAQ keywords, such negative-specific keywords are not defined in advance and by hand. We think such negative-specific keywords can be extracted from negative instances, that is, the past "NOT matching a FAQ" inquiries.

5.2   Keywords Extraction from "NOT Matching a FAQ" Inquiries

Fig. 11 shows how to extract keywords from "NOT matching a FAQ" inquiries. Because "NOT matching a FAQ" inquiries are various, the proposed method cannot identify keywords from all of them. So we propose that keywords of "NOT matching a FAQ" inquiries should be extracted from similar inquiries to the FAQ. First, the proposed method sets words occurring frequently in inquiries matching the FAQ keywords of the FAQ. Second, it calculates Jaccard coefficient of each "NOT matching a FAQ" inquiry sentence and keywords of the FAQ. Finally, it chooses keywords of "NOT matching a FAQ" inquiries from a set of inquires whose Jaccard coefficient is high.

5.3   Creation of Feature Vector of SVM Using Positive and Negative Instances

If the proposed method makes a vector by using extracted keywords, it can reflect features of inquires matching the FAQ and "NOT matching a FAQ" inquiries, however, keywords included in both still exist. These keywords make it difficult to judge whether a inquiry matches a FAQ. Therefore, as shown in Fig. 12, we make characteristic word groups of positive and negative instances except common keywords, and make the characteristic word group of common keywords. As a result, when our system judges whether an inquiry matches FAQ,

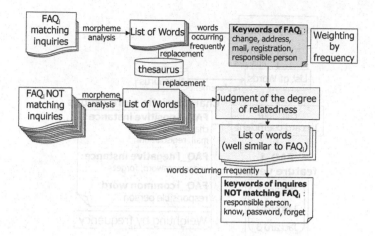

Fig. 11. Extracting keywords from "NOT matching a FAQ" inquiries

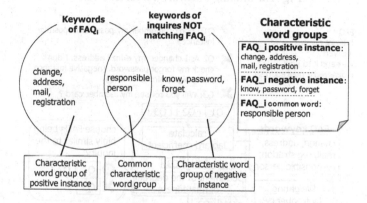

Fig. 12. Creating characteristic word groups

it uses not only words of positive and negative instances but a set of words that make the judgment difficult.

Fig. 13 shows how to make a feature vector of SVM. We calculate Jaccard coefficient of each characteristic word group and an inquiry sentence. By using three Jaccard coefficients as an element of the vector, the feature vector has characteristics of positive and negative instances.

5.4 Removal of NOT Similar Inquiry E-mails

Fig. 14 shows how to remove NOT similar inquiry e-mails. Because keywords of negative instances are selected from the negative instances whose degrees of relatedness to FAQ are high, characteristic word groups do not reflect the features of the negative instances whose degerees of relatedness to FAQ are low. Since SVMs of our system do not judge the inquiries whose degrees of relatedness

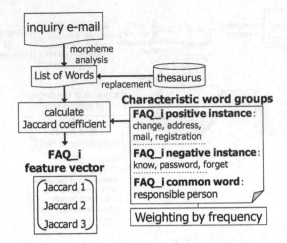

Fig. 13. Making a feature vector of SVM

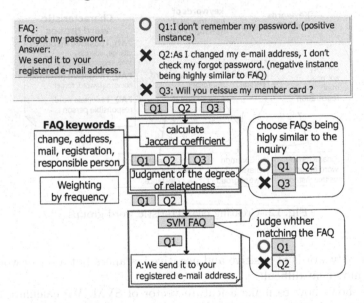

Fig. 14. Removing NOT similar inquiry e-mails

are low, our method separates such inquiries of low relatedness by calculating Jaccard coefficient before processing SVMs.

As a result, SVMs find inquiries which match a FAQ in highly similar inquiries.

5.5   Overview of Identification Method

Fig.15 shows overview of our proposed system. First, it changes an inquiry e-mail into a list of words. Second, it chooses FAQ which is highly similar to the inquiry by Jaccard coefficient. Finally, it determines the matching FAQ by SVM. If it

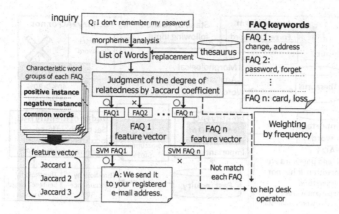

Fig. 15. Overview of identification method

judges that the inquiry does not match each FAQ, it sends the inquiry to help desk operators.

6 A Predictive Search Method of FAQ by Statistical Model of Important Words Co-occurrence

6.1 Outline of a Predictive Search Method of FAQ

When some important words of a certain FAQ are inputted (co-occurred) in the inquiry, the co-occurrence rate of important words differs in each FAQ. Focusing on the difference of the rates, we propose the predictive search method by the rate of important words' co-occurrence in past inquiries. A corresponding FAQ is identified when the co-occurrence rate of important words in the FAQ is the highest. In order to prevent wrong FAQ from being kept displayed, the threshold corresponding to the number of input words is used to decide whether the FAQ is displayed or not. Fig. 16 shows the outline of the predictive search method of FAQ.

The inquiry sentences are processed by morphological analysis to be divided into words, and the synonyms are unified by the thesaurus. Whenever a user inputs a new important word in the inquiry, a corresponding FAQ is identified by the co-occurrence rate of important words in each FAQ. When the co-occurrence rate is larger than the threshold, the FAQ including the important words is considered to be corresponding FAQ to the inquiry.

6.2 Predictive Search by Statistical Model of Important Words Co-occurrence

Fig. 17 shows the outline of the predictive search of FAQ by the statistical model of the co-occurrence of important words. To search FAQ by some of important words, the proposed method generates the statistical model by combination patterns of important words and co-occurrence rates of the important words in past inquiries.

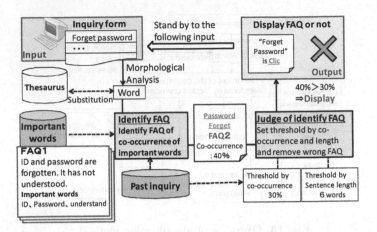

Fig. 16. Outline of FAQ presentation system that uses important word co-occurrence rate model

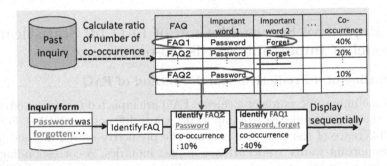

Fig. 17. FAQ search based on stochastic model of co-occurrence of important words

Co-occurrence rates are the ratios of the total number of co-occurrences of important words in past inquiries that correspond to each FAQ or "Other" that means "correspond to no FAQ". The co-occurrences of important words in each FAQi are used for searching each FAQi. Because "Other" have no important words, important words of all FAQ are used for "Other". The *co-occurrence rate*(p, i) of the co-occurrence pattern p belonging to FAQi is obtained in the following expressions:

$$co\text{-}occurrence\,rate(p, i) = \frac{N_{p,i}}{\sum_i N_{p,i}}$$

where $N_{p,i}$ is the number of co-occurrences of p in the past inquiries correspond to FAQi.

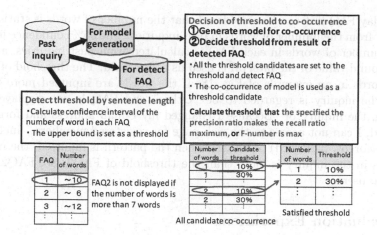

Fig. 18. Decision of threshold for FAQ presentation judgment

As shown in Fig. 17, whenever the co-occurrence pattern p is inputted, FAQi where the *co-occurrence rate(p, i)* is the highest is regarded as corresponding FAQ. FAQ is not displayed when the inquiry is regarded to correspond to "Other". In Fig. 17, when a user inputs "Password", FAQ2 whose co-occurrence rate is 10% is searched. When a user inputs two words "Password" and "forget" additionally, FAQ1 whose co-occurrence rate is 40% is searched.

6.3 Judgment of Displaying FAQ

Fig. 18 shows the outline of deciding thresholds for judgment of displaying searched FAQ. The co-occurrence rate of important words concerning a specific FAQ tends to be increased as the number of input words is increased. In case that the co-occurrence rate is low in spite of a long inquiry, it is possible to judge that the displayed FAQ is wrong. The precision rate can be improved by the threshold for the co-occurrence rate according to the number of words.

First of all, past inquiries are divided for the generation of the stochastic model and for setting threshold. The proposed method calculates co-occurrence rates from past inquiries for the generation of the stochastic model. All the thresholds $Th_{p,i}(W)$ of a co-occurrence pattern p in FAQi are set according to the number of important words W. Changing the candidates of thresholds from 0% to 100% and judging whether FAQ are displayed or not for past inquiries by the thresholds, the proposed method decides thresholds for the system, when the result of the judgment indicates that F-value is maximum or the recall rate is maximum at the certain precision rate.

Moreover, because the average number of words in inquiries depends on which FAQ the inquiry corresponds to, it can be judged whether FAQ should be displayed according to the number of input words. Then, the proposed method does not display FAQ if it can be considered that the number of words is statistically different from the number of words in past inquiries. The 95% confidence interval of the number of words in each FAQ is calculated from past inquiries, and the upper bound value of the interval is used as a threshold. The threshold of FAQ2 is six words in an example of Fig. 18. If the words are inputted more than 7 words, the inquiry is regarded to be "Other" and FAQ is not displayed. For example, the inquiry of "Because I changed my e-mail address and forgot my password, I can not access to the web page for registered members." includes a pattern (forgot, password) of FAQ2. When the pattern is inputted, the number of words in the inquiry is more than the threshold of FAQ2. So, FAQ2 is not shown to users.

7   Evaluation Experiment

7.1   Target of Experiment

In order to evaluate the effectiveness of the proposed system, we perform the experiment that (1)the stepwise clustering method for extracting candidate FAQ, (2)the identification method of inquiry e-mails to the matching FAQ and (3)the predictive search method of FAQ are applied to practical data in a help desk of the member management system of the sports association. Because all the e-mails and FAQ are written in Japanese, Japanese morphological analysis tool "Chasen" [1] is used. How to evaluate each proposed method is as follows:

(1) The stepwise clustering method for extracting candidate FAQ

There are 1318 threads in the target data. Inquiries have 16.8 words and replies have 34.6 words on average (data set I). A constant value (α) in expressions is 0.7 for weighting replies because replies are probably written by particular operators and they use same words in the replies having same contents. The confidence coefficient in adding a thread is 99% and the threshold value in combining clusters is 0.30.

The generated clusters must reflect frequencies of inquiries in input data, and contents of them must be read easily by operators. So, we set the criteria of evaluation as follows:

- Cluster size: The cluster size is defined as the number of threads in the cluster. By comparing cluster sizes each other, we can judge how high the frequency of inquiries is in each cluster, and evaluate which clusters reflect precisely frequencies of inquiries in input data.
- Precision of clustering: The precision of clustering is the rate of threads classified correctly in a cluster. If it is high, operators can read easily a content of a cluster without reading wrong threads.

We compared results of clustering by the proposed method to the conventional hierarchical clustering by Cosine similarity. This conventional method uses the same similarity and clustering method as the proposed method in section 4.2, and a constant value (α) in expressions is also 0.7. The threshold value of the hierarchical clustering is 0.47 that is adjusted to get the best precision manually. We generated clusters by hand and defined the clusters having over 50 threads as candidates of FAQ. Table 1 shows the candidate FAQ and the numbers of threads that have the content of candidate FAQ.

Table 1. Exampeles of candidate FAQ in data set I

	Content	Number of threads
FAQ1	Forgetting my password	210
FAQ2	Correcting my date of birth	123
FAQ3	Altering to player from staff	61

(2) The identification method of inquiry e-mails to the matching FAQ
The number of e-mails is 1845, while 457 of them match to FAQ (data set II). Table 2 shows the number of supervised data and test data. We compared our previous method using SVM[8] to the proposed method in precision and recall rates, by classified inquiry e-mails sent to a help desk of the member management system of the sports association.

Table 2. The number of supervised data and test data (data set II)

FAQ	Q3-3	Q7-2	Q4-7	Q9-8	Q2-1	Q5-9	Q3-4	Q3-8	Q6-7
The number of supervised data	111	27	26	22	13	10	9	7	7
The number of test data	110	26	25	22	12	9	9	6	6

(3) The predictive search method of FAQ
We use 1761 e-mails recorded by operators for a year: while 456 e-mails of them match to FAQ and 1305 e-mails belong to "Other" (data set III). Table 3 shows the number of supervised data and test data. We compared the proposed method to the pattern matching method. e evaluate by the recall rate that corresponding FAQ is displayed once while an inquiry is being inputted, and by the precision rate that FAQ was correctly displayed while inquiry is being inputted. The threshold when F-value becomes the maximum is used as a threshold of the co-occurrence rate. And, the proposed method is applied with confidence levels 95% and 99% for thresholds of the number of words.

Table 3. The number of supervised data and test data (data set III)

FAQ	Q1	Q2	Q3	Q4	Q5	Q6	Q7	Q8
The number of supervised data	13	166	15	9	25	13	35	27
The number of test data	7	83	8	5	12	6	18	14

7.2   Experimental Result

The experimental results by the proposed methods are shown in the following:

(1) The stepwise clustering method for extracting candidate FAQ

Fig. 19 and Table 4 show results of the experiment. They show cluster sizes and precisions of generated clusters having contents of candidate FAQ. In Fig. 19, the sizes of FAQ2 and FAQ3 clusters by hierarchical clustering are almost the same. These clusters do not reflect frequencies of inquiries in input data, which makes it difficult for help desk operators to grasp which content of the cluster is more frequently inquired. On the other hand, the proposed method generates clusters reflecting frequencies of inquiries although there are gaps between cluster sizes of clusters generated by the proposed method and correct clustering. The precisions of FAQ1 and FAQ2 clusters generated by the proposed method are over 70% in Table 4. The precision of FAQ3 cluster by the proposed method is also higher than one by hierarchical clustering. Therefore help desk operators can grasp the contents of the cluster more easily by the proposed method.

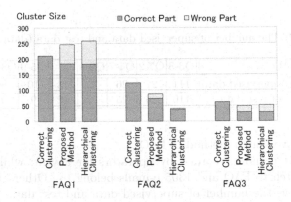

Fig. 19. Result of cluster size

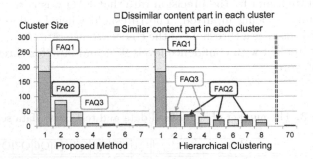

Fig. 20. Clusters generated by each method

Table 4. Result of Precision

Cluster	FAQ1	FAQ2	FAQ3
Proposed method	75%	84%	58%
Hierarchical clustering	71%	90%	55%

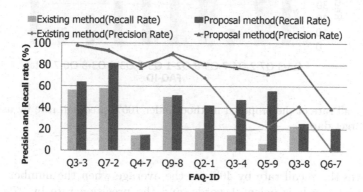

Fig. 21. Result of experiment

Fig. 20 shows a result of all clusters generated by each method. In Fig. 20, the generated clusters are placed in order of the cluster size. Operators extract major clusters as a candidate FAQ by reading all threads in each cluster. And, they judged whether the threads are a similar content part and a dissimilar content part to major threads. The hierarchical clustering generates different clusters even if they have same contents: clusters for FAQ2 and FAQ3 are separated to some clusters. On the other hand, the proposed method does not generate such scattered clusters and generates much less clusters than hierarchical clustering does. So, help desk operators can find candidate FAQ more efficiently by the proposed method.

(2) The identification method of inquiry e-mails to the matching FAQ

Fig. 21 shows the experimental result of the precision and recall rates.

To FAQ whose precision rates are low(from 1% to 40%) by using the previous method, they are higher(from 40% to 80%) by this proposed method. In addition, Recall rates of inquiries of the most and second most numbers are also higher. However, 100% of recall rate of target classification data is not guaranteed as far as the system adjusts the threshold by using learning data. If we can adjust the threshold to the value which makes precision rate of classification data 100%, our system classifies in 50% recall rate, as shown in Fig. 22. Therefore, the future challenge is to setting such threshold automatically.

(3) The predictive search method of FAQ

Fig. 23 shows the experiment result of precision rates and recall rates in applying the pattern matching and the propoased method with confidence levels 95% and 99%. The proposed method with the confidence level 95%

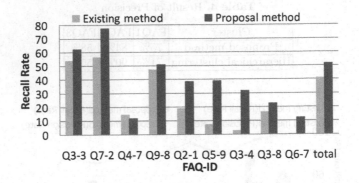

Fig. 22. Recall rate of the proposed method under 100% precision rate by using classification target data

improves the recall rate by 40% on the average when the number of input important words is below 2 words, and the precision rate by 27% on the average when the number of input important words is more than 3 words. In applying the method with pattern matching to long inquiries that have important words in some FAQ, wrong FAQ is often displayed.

Next, we discuss the effect of the confidence level for setting thresholds. As shown in Fig. 23, the difference of recall rates or precision rates between results with both of the confidence levels is below 10%. Because the difference of the confidence levels does not make the effect on the recall rates or precision rates, operators can set the confidence level without strictly adjusting. In this help desk, the operators want to keep the recall rate higher in inputting few words and the precision rate higher in inputting more words. So, the confidence interval 95% is appropriate for this help desk.

We show examples of inquiries where the proposed method improves the recall rate and the precision rate in the following:

- Inquiry A "Though I want to pay, the time limit of payment had passed the date yesterday..."
  In this inquiry, it is necessary to input an important word "payment", "time limit" and "pass". But, in the proposed method, corresponding FAQ can be displayed by inputting just "pass" and "time limit".
- Inquiry B "I don't know my membership code. And, the time-limit of payment ..."
  Because this inquiry includes two or more questions and is longer, it must be classified correctly into "Other". The threshold of the number of words can prevent wrong FAQ from being displayed.

Through the above discussion, it is confirmed that the proposed method is effective for the predictive search of FAQ. Now, operators must decide important words used by the proposed method. So, our future task is to decide the appropriate important words by the past inquiries.

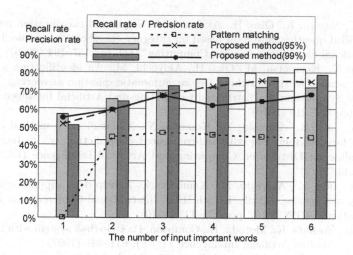

Fig. 23. Result of experiment by the predictive search method

8 Conclusion

We proposed the help desk support system based on relationship between inquiries and responses. By using the relationships of the similarity, exclusivity and co-occurrency between an inquiry and a response, the proposed system provides the stepwise clustering method for extracting candidate FAQ, the identification method of inquiry e-mails to the matching FAQ and the predictive search method of FAQ. Applying the proposed methods to practical data in the help desk, we confirmed that the proposed system with the methods are effective enough to support the help desk.

References

1. Chasen, http://chasen-legacy.sourceforge.jp/
2. Cortes, C., Vapnik, V.: Support-vector networks. Machine Learning 20(3), 273–297 (1995)
3. Foo, S., Hui, S., Leong, P., Liu, S.: Integrated help desk support for customer services over the world wide web - a case study. Computers in Industry 41(2), 129–145 (2000)
4. Halverson, C., Erickson, T., Ackerman, M.: Behind the help desk: Evolution of a knowledge management system in a large organization. In: ACM Conference on Computer Supported Cooperative Work, pp. 304–313 (2004)
5. Hammond, K., Burke, R., Martin, C., Lytinen, S.: Faq finder: a case-based approach to knowledge navigation. In: 11th Conference on Artificial Intelligence for Applications, pp. 80–86 (1995)
6. Hsu, C.H., Guo, S., Chen, R.C., Dai, S.K.: Using domain ontology to implement a frequently asked questions system. In: World Congress on Computer Science and Information Engineering, vol. 4, pp. 714–718 (2009)

7. Iida, K., Negoro, K., Oiso, H., Akiyoshi, M.: Sequential classification method for open-ended questionnaire data based on category classification sample. In: IADIS International Conference Applied Computing 2009, pp. 157–164 (2009)
8. Itakura, K., Kenmotsu, M., Oka, H., Akiyoshi, M.: An identification method of inquiry e-mails to the matching faq for automatic question answering. In: International Symposium on Distributed Computing and Artificial Intelligence (DCAI 2010), pp. 213–219 (2010)
9. Iwai, K., Iida, K., Akiyoshi, M., Komoda, N.: A classification method of inquiry e-mails for describing faq with self-configured class dictionary. In: International Symposium on Distributed Computing and Artificial Intelligence (DCAI 2010), pp. 35–43 (2010)
10. Iwai, K., Iida, K., Akiyoshi, M., Komoda, N.: A help desk support system with filtering and reusing e-mails. In: 8th IEEE International Conference on Industrial Informatics (INDIN 2010), pp. 321–325 (2010)
11. Kang, B., Yoshida, K., Motoda, H., Compton, P.: Help desk system with intelligent interface. Applied Artificial Intelligence 11(7-8), 611–631 (1997)
12. Kim, H., Seo, J.: Cluster-based faq retrieval using latent term weights. IEEE Intelligent Systems 23(2), 58–65 (2008)
13. Muller, N.: Expanding the help desk through the world wide web. Information Systems Management 13(3), 37–44 (1996)
14. Salton, G., McGill, M.J.: Introduction to Modern Information Retrieval. McGraw-Hill (1983)
15. Salton, G., McGill, M.J.: Implementing agglomerative hierarchical clustering algorithms for use in document retrieval. Information Processing and Management 22, 465–476 (1986)
16. Sneiders, E.: Automated faq answering with question-specific knowledge representation for web self-service. In: 2nd Conference on Human System Interactions (HSI 2009), pp. 298–305 (2009)
17. Sullivan, D.: Document Warehousing and Text Mining: Techniques for Improving Business Operations, Marketing, and Sales. John Wiley and Sons (2001)
18. Tsuda, Y., Akiyoshi, M., Samejima, M., Oka, H.: A classification method of inquiry e-mails for describing faq with automatic setting mechanism of judgment threshold values. In: International Conference on Enterprise Information Systems (ICEIS 2012), vol. 3, pp. 199–205 (1998)
19. Vapnik, V.N.: The Nature of Statistical Learning Theory. Springer (1995)
20. Willett, P.: Recent trends in hierarchic document clustering: A critical review. Information Processing and Management 24(5), 577–597 (1988)
21. Yang, S.Y.: Developing an ontological faq system with faq processing and ranking techniques for ubiquitous services. In: First IEEE International Conference on Ubi-Media Computing, pp. 541–546 (2008)
22. Zamir, O., Etzioni, O.: Web document clustering: A feasibility demonstration. In: 21st Annual International ACM SIGIR Conference on Research and Development in Information Retrieval, pp. 46–54 (1998)

The Impact of Advertising in Emergent Economic Context: A System Dynamics Simulation Approach

Cuauhtémoc Sánchez Ramírez, Guillermo Cortés Robles, and Giner Alor Hernández

Postgraduate Departement. Instituto Tecnológico de Orizaba, Av. Oriente 9,
No. 852. Orizaba Veracruz, México
{csanchez,gcortes,galor}@itorizaba.edu.mx

Abstract. The continual changes in the current market have increased complexity in business strategies. Under such dynamic conditions, enterprises need new tools, methods and methodologies to deal with uncertainty, ambiguity and a huge number of interactions. This phenomenon presents important differences in emergent economies but also similarities: the success of a product in the market will depend on product design, an efficient supply chain and also upon the right advertising campaigns in order to improve profit. In parallel, enterprises in emerging economies are involved in a global business where it is necessary to consider the challenging conditions for better decision-making. System dynamics simulation is a tool that can helps enterprises to analyze and observe the effect of business strategies over global performance. Marketing and advertising considered as a business strategy is not an exception to this statement. Thus, in this article is described a methodology and a study case where system dynamics simulation helps a motorcycle distributor enterprise to assess the impact of their advertising strategies in the market share, in order to chose the most suitable strategy through a sensitivity analysis.

Keywords: Advertising, System Dynamics, Emergent Economy, Marketing, Simulation.

1   Introduction

Marketing is an organizational activity where several human abilities and technical capacities are combined to achieve performance. It is also the process to deliver something important or valuable between buyer and seller. For this reason, marketing is usually considered as an exchange process where some individuals obtain something they need or want and others develop goods, services or any other combination of tangible or intangible characteristics. This exchange is thus essential in any open economic system. In recent years the concept of marketing has suffered radical transformations. Its original business-centered perception has change to a wider concept: one where all organizations interact in a global and interconnected environment, including non-profit organizations such as hospitals, universities, religion communities or any other community of practice.

T. Matsuo & R. Colomo-Palacios (Eds.): *Electronic Business and Marketing*, SCI 484, pp. 67–83.
DOI: 10.1007/978-3-642-37932-1_6   © Springer-Verlag Berlin Heidelberg 2013

According to this conception, the American Marketing Association (AMA) underlines that marketing has evolve as "...*the process of planning and executing the conception pricing, promotion, and distribution of ideas goods, and services to create exchange that satisfy individual and organizational goals*". Marketing is also seeing as "...*the activity, set of institutions, and processes for creating, communicating, delivering, and exchanging offerings that have value for customers, clients, partners, and society at large*" [2].

The marketplace is then a very complex environment, which is continually changing and demanding new organizational capacities. This dynamism affects the way marketing is managed, increasing connections among apparently unrelated stakeholders, demanding new abilities for revealing what costumer wants and needs, and considerably reducing the time for developing new products and services [6]. Under this new marketing perspective, new tools and approaches are needed in order to face new challenges. Farris et al [10] focus its attention in one of the mains issues of marketing: assessing the impact of any marketing strategy in order to evaluate performance, to anticipate obstacles and to observe what will be the possible market reaction. This article considers that any marketing strategy is embedded in a system and for this reason, that several complex interactions occur that could be explored through a system approach: system dynamics simulation.

This article integrates two identified tendencies: to evaluate productivity considering the most important marketing metrics by means of system dynamics. The objective is to analyze the impact of advertising in one emergent market-place using system dynamics simulation. The article is organized as follows: first section describes the state of the art from different perspectives (i.e. marketing research, system dynamics simulation, management). Next section briefly describes the methodology and the model assumptions. In the same section is also defined the context for developing a system dynamics model simulation. In third section summarizes results and discussion for finally, in the last section to delineate future research.

2   State of the Art

According to Varey [37] marketing is an interaction system that could be defined through relationship, interaction and useful networking. This interactive marketing involves not only customers but also competitors and even other domains. Then, marketing is a social interaction based on communication and functions/processes that interact continually and evolve in this exchange. Following this conception, market interaction should be considered as a "*complex dynamic adaptive interaction system*" that must be analyzed taking into a count all relationships and interactions.

Arinze and Burton (1992) underline the importance of simulation for developing an effective marketing strategy. Authors point out that inadequate data sources, uncertainty, and complex interaction difficult decision-making. This problem could be partially solved through a simulation model because it is possible to represent the stochastic comportment of most of the component in interactive marketing. Authors utilize a simulation model as a tool for marketing decision support process. This model employs Monte Carlo simulation to represent market dynamics, comparative marketing and competitive actions to determine performance. Even if this model is useful, it does not contemplate other interactions as competitors, new products in the market to mention some of the most important elements.

Otto [28] describes a process useful to test the effectiveness of different market entry strategies, and to explain the intrinsic logic of a system. This process is modeled with a system dynamics view, which integrates the opinion and experience of an expert panel. According to the author, this process increases the sense of ownership in users and facilitates adoption. Nevertheless, the model is specific to market entry strategies and could not be suitable to explain past strategies. This model could be useful to predict tendencies when launching new strategies, which is one of its most important advantages.

Chan et al. [7] propose a system dynamics model for customer relationship management (CRM) using iThink®. In this model are integrated the customer purchasing behavior, a Markov chain and some financial indicators. These elements are combined to calculate the probability of costumer's second time purchase. This data is later incorporated to another module where is estimated the customer lifetime value. Finally, this information enables the model for calculating the return of investment in the long term. The model was conceived to evaluate the business strategy according two perspectives: marketing strategies and product development. The combination of both perspectives creates the condition to evaluate what should be the parameter to deploy the best marketing strategy that reach maximum profit and also maximum value for stakeholders. This model anticipates the impact that product development and marketing strategic decision could have for an organization, revealing the need for more adapted models, especially for really changing environments.

Following this reflection, Leeflang and Wittink [24] established that in today enterprises or environment that change rapidly, the need for marketing models that could be exploited in different contexts is increasing, impelling the development of new and different modeling approaches. This article emphasizes the necessity to integrate financial parameters, creativity and marketplace tendencies to implement successful marketing strategies.

Größler et al., [17] explain in a more specific situation, that price reductions and/or product enhancements -usually seen as appropriate ways to increase market share-, combined inadequately can lead to substantial loss in sales revenue. The author employs a system dynamics model to analyze the price and product strategies in the capital goods industry. The proposed model is useful to formalize some intuitive beliefs and also to observe the impact of different marketing strategies. Author established that the model had a significant impact for an organization because offers a wide perspective about mechanisms of markets and competition.

Wierenga and Ophuis [39] propose a model to deal with marketing decision support systems (MDSS) based on five categories of factors that affect adoption, use, and satisfaction: external environment factors, organizational factors, task environment factors, user factors and implementation factors. The main hypothesis was tested on data obtained from a survey of 525 companies. Among the main conclusions of the article are: (1) adoption of a marketing decision support system does not depend on the same factors that determine success, once the system deployed. (2) Communication systems and knowledge from inside the company are essential for adopting this kind of systems. (3) Success of the system will need a proactive participation of user, an adaptive capacity to different context and requirements and also, the ability to integrate to high-level systems. Maybe the most important issue discussed in this article is that marketing decision support systems are adopted with the aim to acquire valuable information, not to added value to existing information. Finally, this article offers some basic parameter useful to conceive a marketing decision support system with the goal to increase users adoption and satisfaction, but it does not considered all factors simultaneously, condition that could produce some tendencies according the expert's technical domain.

Sheth and Sisodia [34] note a low productivity in marketing strategies –with an associated difficulty to measure it- and also the lack of understandable indicators. Authors suggest that productivity could be improved if new marketing strategies turn their attention from markets to individual customers and also if marketing shift the way is managed: from a budget expensed annually to a function that could be amortized over time. This article highlights a very relevant problem for marketing strategies: the partial absence of adequate metrics to evaluate productivity.

Another approach needs to be described in this article: system dynamics (SD). The subjacent reason is that system dynamics enables managers and marketing thinkers to observe the impact of one parameter in all the system. SD is then useful to build a coherent representation of interactions within a system [32].

2.1   System Dynamics (SD)

SD is a relatively new research field. Its application domain is expanded continuously, due to its versatility and application in non-technical domains. This diversity creates different perceptions about this approach. Next points briefly describe different points of view:

SD is a methodology useful to represent and simulate system complex. This goal can be achieved through the analysis of causal relationships among the components that integrates a system. Thus, SD offers a way to understand what are the conditions and internal mechanisms that produce certain behavior in a system, or the relationship that exists between the system structure and its behavior. It is then, a modeling and simulation methodological approach that can helps individual to understand systems over time, involving several stakeholders, uncertainty, complexity and not clear defined causal relationships [5].

A computer-aided approach for analyzing and solving complex problems with a focus on policy analysis and design [3][36].

An approach that uses a perspective based on information feedback and delays to understand the dynamic behavior of complex physical, biological, and social systems. *"… the study of the information feedback characteristics of industrial activity to show how organizational structure, amplification (in policies), and time delays (in decision and actions) interact to influence the success of the enterprise"* [12].

2.2   SD Main Concepts

Modeling systems demand a creative effort and some essential concepts to produce schemas that could be understood in different contexts. Next points describe some of the most important SD concepts:

- A feedback loop is a concatenation of causes and effects that produces over time, a change in a given variable. Before this first impact, a countercurrent impact travels around the loop and comes back to affect the initiating variable. If it is possible to observe a change (increase or decrease) on the same variable in the same direction, then the feedback loop is identified as a positive (self-reinforcing) feedback loop. If an initial increase (or decrease) of a variable creates an effect on the same variable in the opposite direction, then the feedback loop is identified as a negative (balancing) feedback loop [4].
- A time delay could be observed in a process where the output lags behind its input.
- A causal loop diagram represents the feedback, time delays and nonlinearities between variables [30].

As mentioned before, SD has been applied in technical domains such as dynamic decision making [35], [40], supply chain management [33], complex non-linear dynamics [26], software engineering [1][8][22], to mention a few, and due to its nature, DS can be also applied in non-technical domains. As an example: Corporate planning and policy design [12], public management and policy [20], biological and medical modeling [18], Chinese medicine [19], environmental issues [11], sociology [29], theory development in the natural and social sciences [9],

2.3   Discussion

Marketing is a complex domain that needs a more comprehensive approach to explain ambiguous interaction, to involve uncertainty and most of all; an approach capable to deal with all this conditions simultaneously and continually evolving in time. Next table summarize several articles that aim to solve this problem.

Table 1. Comparison among different authors and topics in marketing and SD

Author	Applied in marketing	Evaluates productivity	Marketing metrics	Applied in emergent markets	Simulates scenarios
1	✔	✘	✔	✘	✘
2	✔	✘	✔	✘	✘
3	✔	✘	✔	✘	✔
4	✔	✔	✔	✘	✔
5	✔	✘	✔	✘	✘
6	✔	✔	✔	✘	✔
7	✔	✘	✔	✘	✔
8	✔	✔	✔	✘	✘

1) Farris et al., 2010; 2) Varey, 2008; 3) Otto, 2008; 4) Chan et al., 2010; 5) Leeflang et al., 2000; 6) Größler et al., 2008; 7) Wierenga et al., 1997; 8) Sheth et al., 2002.

This concise analysis shows that the evaluation of productivity needs to be explicitly considered while proposing new marketing strategies. Even if almost all articles uses at least one indicator, there is no a consistent use of marketing metrics. Maybe the most important partial conclusion of this survey is that there is not any model applied in emergent markets or economies. Finally, but not least important, in the complex interaction of several marketing factors, the ability to simulate different scenarios is an undeniable advantage.

According to this conclusion, next section describes how was developed a simulation model to analyzed the campaign to advertising and used this model to help an organization to define the most appropriated marketing strategy in an emergent economy.

3   Applying SD in Advertising and Marketing

In development of marketing strategies, the companies need to use tools and methodologies that help them to comprehensively describe the impact of their decisions in areas such as: market, income customers, among others. Due to the complexity and dynamism of the factors and elements that must be considered in a marketing strategy, a tool that helps to analyze these factors is simulation. Among the main simulation advantages are: it allows a middle position between pure formal modeling, empirical observations and strategies used to improve decisions [16], it also provides the ability to include estimations about soft factors that are not easy to measure.

In this article the simulation methodology that it used to analyze complex systems in business administration and marketing is the system dynamics [31] because it can managed uncertainty, ambiguity and a vast quantity of variables. System dynamics was development by Jay Forrester [12] to show how decisions, policies, system structures, and delays interact to influence growth and stability. A system dynamics model captures the multiple feedback loops underlying the endogenous behavior of a particular problem.

3.1   Methodology

System dynamics is typically employed in two generic situations: (1) to identify the dynamic relationships, and then try to understand what could be the possible consequences if one or any combinations of these relationships are materialized. This analysis is carried out linking observation with some methods, models or techniques that could help in this process of validate one or a set of hypotheses. (2) To simulate the dynamic relationships in order to reveal what could be the result if different intensity, power, amount or a combination of parameters (time, delay and feedback) are modified. In both cases, a generic methodology could be used.

1. Identify a problem and advance at least one dynamic hypothesis to explain what is the cause of a problem
2. Build a model of the system
3. Validates the model, which means to observe if the model really represents the real world
4. Modify conditions to observe how this change reflects into the system and then reformulate hypotheses if necessary
5. Analyze results and conclude [12], and [35].

3.2   A Case Study: Exploring Relationships in Emergent Market with DS

The case study is about a motorcycles distributor -called ABC enterprise in this document- located in the state of Veracruz, Mexico. This enterprise has seen a substantial reduction in their main financial indicators thus; a new strategy should be deployed to revert this condition. The enterprise needs to answer several questions: What will be the impact of advertising? What will be the right budget? What will be the right time horizon for this strategy? What will be the impact in the enterprise's key financial indicators? To answer all this questions, a system dynamics simulation needs to be conceived. Next paragraphs describe how the methodology described above is applied.

3.3   Describing the Problem

The evaluation of ABC enterprise financial and sales performance indicators show that its market share has declined. Before an interview with salesmen, managers and client service department, this is a probably consequence of new competitors in the market. A particular competitor offers some products with a considerably low price, for a slightly different market, a market for a lower income. Therefore, the manager wants to analyze the impact of advertising in sales and to see what the relationship is over time of this strategy in order to assign the right budget.

3.4   Causal Loop Diagram

Next step of modeling and simulation with system dynamic is to develop a causal loop diagram to analyze graphically, the relationship between model variables and feedback loops. According to Georgiadis et al [15], *"the negative feedback loop exhibits a goal-seeking behavior after disturbance; the system seeks to return to an equilibrium situation. In a positive feedback loop an initial disturbance leads to further change, suggesting the presence of an unstable equilibrium"*.

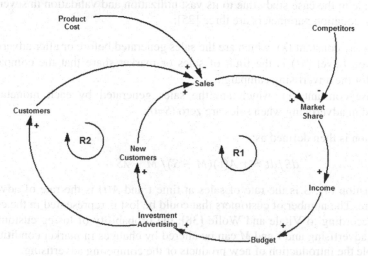

Fig. 1. Causal loop diagram of advertising in distributor of motorcycle

To develop the causal loop diagram, interviews were conducted with the manager and other stakeholders in the enterprise. As a result of these interviews, nine key variables were identified: customers and new expected costumers, product cost or market price, sales level, competitors, market share, income, budget and advertising investment (see figure 1). Several previous works served as a theoretical foundation for the development of the model schematized in figure 1 [10][21][27][38].

One of the most important characteristics in causal diagrams are feedback loops, thus next points describe most relevant feedback loops.

Loop R1. This loop represents the sales impact that can be obtained through advertising, especially for attracting new customers. If sales increase, then it will have a positive influence over the market. However, competitors negatively influence the market share, which means that if the numbers of competitors increase in the market then the share will decrease. Otherwise, if the enterprise´s participation increases, then the sales will increase and the budget for advertising too. The risk of this reinforcing loop is that if sales decline, income will follow this negative tendency affecting in the same direction budget and market share.

Loop R2. This loop is similar to loop R1, but in this loop is analyzed the impact of sales in regular customers. In this case, it is possible to observe that if the price to market gets increased, then the sales will decrease, revealing that competitors are an exogenous variable of the model.

3.5   Equations of the System and Initial Validation

Huang et al [21], present a state of the art about a variety of differential equations that have been used to analyze investment in business advertising. Among the most important equations there is one proposed by Vidale and Wolfe [38]. This equation plays an essential role in the case study due to its vast utilization and validation in several situations. The equation parameters are three [25]:

1. Sales Decay constant (λ), which are the sales generated before or after advertising
2. Saturation Level (M) is the limit of sales or market share that the company can obtain for the advertising campaign
3. Response constant (r), which are the sales generated by each monetary unit invested in advertising when sales are zero ($S=0$).

The equation is then defined as:

$$dS/dt = r\,A(t)(M - S)/M - \lambda S \tag{1}$$

In the equation (1), S is the rate of sales at time t and $A(t)$ is the rate of advertising expenditure. The number of customers that could be lost is represented in the equation for λS. According to Vitale and Wolfe [38], the probability of losing customers decrease by advertising and r and M can be altered by changes in market conditions (i.e. for example the introduction of new products or the competing advertising.

In this case study, a sensibility analysis was carried out by changing the value of r and M to observe the impact on sales. If the advertising level is maintained for a time T, the rate of sales is obtained by integrating the equation (1) as:

$$S(t) = [M/(1 + \lambda M/rA)]\{1 - e^{-(rA/M+\lambda)t}\} + S_0 e^{-r(A/M+\lambda)t} \ , (t < T) \tag{2}$$

The sales in t=0 is represented in equation (2) for S_0. When the advertising campaign has finished (t> T), the sales decrease exponentially (equation 3).

$$S(t) = S(T)e^{-\lambda(t-T)}, \quad (t > T) \tag{3}$$

It is possible to measure the gain of market share for the enterprise considering its revenues. The equations 4 proposed by Farris et al [10] have proved its utility:

$$Market\ Share\ (\%) = \frac{Sales\ Revenue}{Total\ Market\ Revenue} \tag{4}$$

According to Farris et al [10], *"market share is an indicator of how well a firm is doing against its competitors and helps managers evaluate both primary and selective demand in their market"*. This equation was also used as a validation strategy. Past advertising campaigns were the basis to build an approximate comparison. It is important to note that there are several factors affecting advertising productivity, some of them are easily measurable but other are subjective (i.e. creativity in advertising, beauty of the product, among others).

3.6   Analysis and Results

This section describes the results obtained before running the simulation model represented in figure 1. The data used to verify and validate the model come from the ABC enterprise. Nevertheless, because of the standard structure of the mathematical model, it conclusion can be generalized to other companies from the same sector. The simulation model was launched in the software STELLA®, with a simulation length of 12 months.

The values used in this model (Table 2) were obtained in the ABC enterprise. The estimated total market of motorcycles in the region is approximately of $250,000 dollars monthly. The enterprise has an average sale of $25,000 dollars, the saturation

Table 2. Variables and values of the model

Variables	Values
Estimated Total Market	$ 125,000 dollars (Monthly)
Average Monthly Sales (S) *	$ 25,000 dollars (Mo)
Saturation Level (M)	$ 55,000 dollars (Mo)
Advertising Investment	$ 10,000 dollars (Mo)
Response Constant (r)	2.31 (Mo)
Sales Decay Constant (λ)	0.9 (Mo)
Advertising Campaign Months *	6
Total Competitors	4

*Initial values.

level of their sales is estimated in $55,000 dollars, the advertising investment can not exceed $ 10,000 per month, and the sales decay are estimated in approximated 9% monthly.

Figure 2 shows the expected sales using available data (see table 2). It is possible to observe that while the advertising campaign is on the market, sales increased exponentially. Following the model, when advertising stops, the sales decrease exponentially too. In this scenario the sales begin with $ 25,000 dollars, the monthly average. The maximum value was $ 52.359 (sixth month) and the average income in twelve month was $ 30.577 dollars.

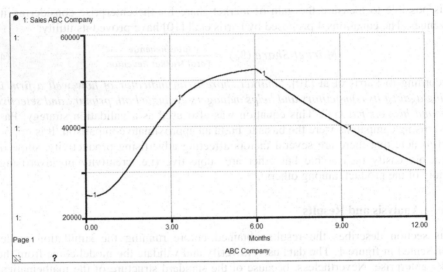

Fig. 2. Total sales with six months of advertising

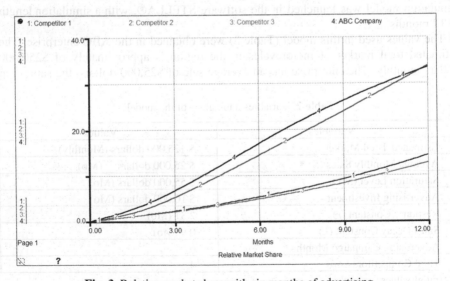

Fig. 3. Relative market share with six months of advertising

Although the sales of the ABC enterprise increase as a consequence of advertising, the market share was less than expected, especially when comparing the enterprise and competitor number 3 (Figure 3).

The conditions experimented and observed in figure 3, suggests that additional measures need to be taken. Consequently, next section describes different scenarios with the aim to validate the model and to generate alternatives to increase the ABC enterprise market share.

3.7   Model Validation and Sensitivity Analysis

Validation in the system dynamics methodology is one of the most important steps. The goal in this step is to verify that the uncertainty level is adequate under the model assumptions. If the model passes the test, then it could be used to analyze different scenarios or to verify policies. The causal loop diagram is the first element to validate the model, because in this schema is possible to identify the influences among all system components. Forrester & Senge [13], and Kleinjen [23] suggest an excellent way to validate a model: to conduct a sensitivity analysis.

Table 2. Values for the sensitivity analysis

Variable	Values (Months)	Total Sales (Year)	Relative Market Share (Year)
Months	0	$ 183,446	12.2%
	3	$ 393,625	26.2%
	6	$ 502,292	33.5%
	9	$ 557,868	37.2%

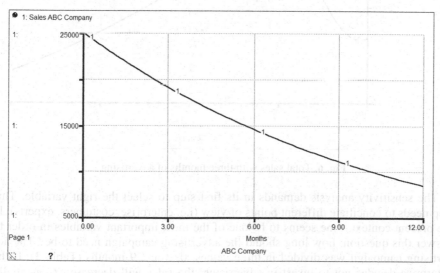

Fig. 4. Total sales with zero investment of advertising

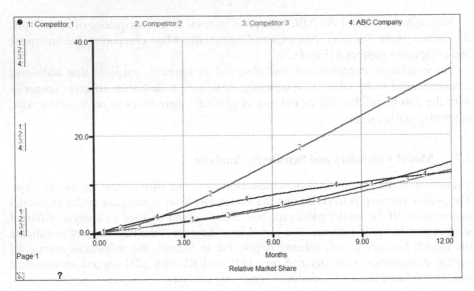

Fig. 5. Relative market share with zero investment of advertising

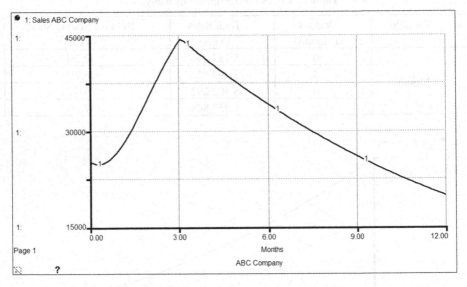

Fig. 6. Total sales with three months of advertising

The sensitivity analysis demands in its first step to select the right variable. This step needs to conciliate different points of view (i.e. enterprise, costumers, experts). In the present context, time seems to be one of the most important variables in order to answer this question: how long should the advertising campaign need to be? The advertising campaign was divided in three values, 0, 3 and 9 months (Table 3). If the enterprise decides not to invest in advertising, the sales will decrease exponentially (Figure 4) at the end of twelve months the total sales would be about $183,446 and the market share about 12.2% (Figure 5).

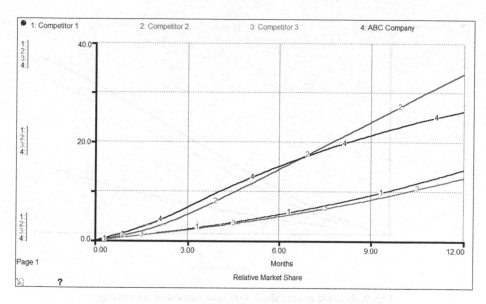

Fig. 7. Relative market share with three months of advertising

If the advertising campaign remains for three months, then the sales at the end of the year will be $393,625 and the market will reach 26.2% (Figure 6 and 7 respectively).

In another scenario, if the distributor decided to invest for nine months, the total sales would be $557,869 with 37.2%, exceeding their competitors (Figure 8 and 9).

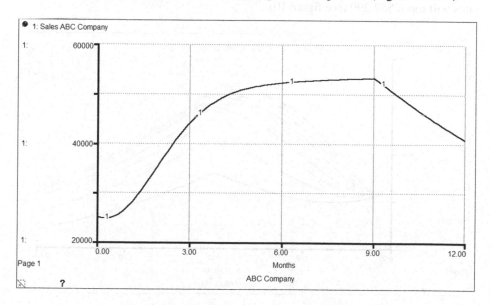

Fig. 8. Total sales with nine months of advertising

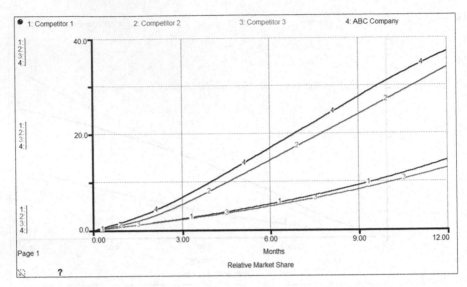

Fig. 9. Relative market share with nine months of advertising

It is necessary to answer another important question: what will be the impact if the monthly budget for advertising changes? A sensitivity analysis to modify this parameter from $1,000 to $10,000 reveals that the investment has a positive impact over sales and market share (more investment, more sales). As an example: if advertising investment is $1000, sales can reach $29,765. If this investment reaches $10,000, then sales will touch $52,299 (see figure 10).

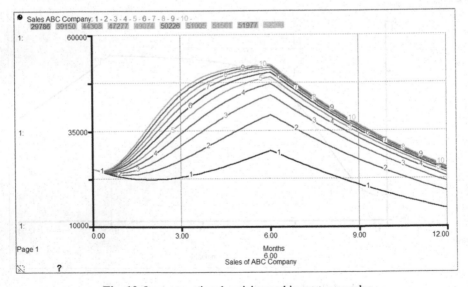

Fig. 10. Investment in advertising and impact over sales

Apparently, the right level of investment is around $5,000; however, it is necessary to consider the total market share. This means that $5,000 will reduce the total market share compared with $10,000 of advertising investment, but will have a satisfactory impact over sales, thus the enterprise must decide what should be the investment taking into account the total market share.

4   Conclusions and Future Research

The contribution of this article is the creation of a system dynamics simulation model useful to analyze the impact of advertising in the context of an emergent market. This particular context is characterized by complex phenomena such as markets in continual expansion that can accept new products at higher rates; a costumer's income that could be classified in both extremes of the economic and social scale; a market with fuzzy regulations that allows the entry of products that not always meet the basic requirements established for other countries, for mention a few. Thus, the conditions of this economies demand new tools, methodologies and techniques to face problems satisfactorily in order to increase industrial and commercial performance, but also useful to assist the managerial decision process. System dynamics simulation and its inherent methodology had proved its usefulness to deal with these issues in technical and non-technical domains.

The study case described in this article explains how to apply the system dynamics methodology in order to build a simulation model. The mathematical and theoretical foundations of this model where described with the aim to sustain decision taking and to define the most appropriated advertising campaign (considering budgetary restrictions and seeking the highest possible benefit over time). The causal diagram of the model shows that two feedback loops define the importance of investment over sales, revealing the exogenous variables of the model. Likewise, the equation proposed for Vidale and Wolfe [38], supports the mathematical structure of the model. This structure enables a sensitivity analysis when different parametric changes affect the equation, helpful condition that can guide creativity to propose a set of scenarios. The objective is then to test each scenario or a combination that would be the most adequate under the market conditions that restraint the ABC enterprise to recommend the conditions that will increase sales and the market share. The best decision in the emergent market where the ABC enterprise is immersed points out that with the available budged the best scenario is to conserve the advertising campaign during nine months, expecting total sales per year of $557, 868 dollars and 37.2 % of the market share, which is a really significant improvement.

The nature and conditions of the case study open a really important research domain and future work: to analyze the impact of advertising in supply chains, because the relationship among the huge diversity of elements in the supply chain is a complex system. Although there are some studies reported, there are significant research opportunities if advertising, marketing metrics and simulation is integrated within the supply chain. Another emergent research filed is to analyze with system dynamics, the impact on the market due to new product development activities combined with the supply chain [14].

References

1. Abdel-Hamid, T.K.: The Dynamics of Software Development Project Management: An Integrative System Dynamics Perspective. PhD Thesis, Sloan School of Management, MIT, Cambridge, MA (1984)
2. AMA- American Marketing Association, definition of marketing, http://www.marketingpower.com/_layouts/Dictionary.aspx?dLetter=M#marketing+management (consulted in August 2012)
3. Angerhofer, B., Angelides, M.: System dynamics modeling in supply chain management: Research review. In: Proceedings of the 2000 Winter Simulation Conference (2000)
4. Besiou, M., Stapleton, O., Wassenhove, L.: System dynamics for humanitarian operations. Journal of Humanitarian Logistics and Supply Chain Management 1(1), 78–103 (2011)
5. Brian, F.: Analytical System Dynamics. Springer (2009)
6. Carayannis, E., Campbell, D.: Editorial preface to the first volume of Journal of Innovation and Entrepreneurship. Journal of Innovation and Entrepreneurship 1(1) (2012), doi:10.1186/2192-5372-1-1
7. Chan, S., Ip, W., Cho, V.: A model for predicting customer value from perspectives of product attractiveness and marketing strategy. Expert System with Applications 37(2), 1207–1215 (2010)
8. Colomo-Palacios, R., García-Crespo, A., Gómez-Berbís, J.M., Paniagua-Martín, F.: A Case of System Dynamics Education in Software Engineering Courses. IEEE Multidisciplinary Engineering Education Magazine 3(2), 52–59 (2008)
9. Dill, M.: Capital Investment Cycles: A System Dynamics Modeling Approach to Social Theory Development. Paper read at 15th International System Dynamics Conference: "Systems Approach to Learning and Education into the 21st Century", Istanbul, Turkey (1997)
10. Farris, P., Bendle, N., Pfeifer, P., Reibstein, D.: Marketing Metrics. The definitive guide to measuring marketing performance, 2nd edn. Pearson Educations Inc. (2010)
11. Ford, A., Lorber, H.W.: Methodology for the Analysis of the Impacts of Electric Power Production in the West. Paper read at Environmental Protection Agency Conference on Energy/Environment II (1977)
12. Forrester, J.: Industrial Dynamics. Productivity Press, Portland (1961)
13. Forrester, J., Senge, P.: Test for Building Confidence in System Dynamics Models. TIMS Studies in the Management Science 14, 209–228 (1980)
14. Galanakis, K.: Innovation process. Make sense using systems thinking. Technovation 26(11), 1222–1232 (2006)
15. Georgiadis, P., Vlachos, D., Iakovou, E.: A system dynamics modeling framework for the strategic supply chain management of food chain. Journal of Food Engineering (70), 351–364 (2005)
16. Größler, A., Schieritz, N.: Of stocks, flows, agents and rules strategic simulation in supply chain research. Research methodologies in supply chain management, pp. 445–460. Editorial Physica-Verlang (2005)
17. Größler, A., Löpsinger, T., Stotz, M., Wörner, H.: Analyzing price and product strategies with a comprehensive system dynamics model—A case study from the capital goods industry. Journal of Business Research 61(11), 1136–1142 (2008)
18. Hansen, J.E., Bie, P.: Distribution of body fluids, plasma protein, and sodium in dogs: a system dynamics model. System Dynamics Review 3(2), 116–135 (1987)
19. Herfel, W., Gao, Y., Rodrígues, D.: Chinese Medicine and Complex Systems Dynamics. Philosophy of Complex Systems 1, 675–719 (2011)

20. Homer, J.B., Clair, C.L.: A Model of HIV Transmission through Needle Sharing. A model useful in analyzing public policies, such as a needle cleaning campaign. Interfaces 21(3), 26–29 (1991)
21. Huang, J., Leng, M., Liang, L.: Recent developments in dynamic advertising research. European Journal of Operational Research 220(3), 591–609 (2012)
22. Kahen, G., Lehman, M., Ramil, P., Wernick, P.: System dynamics modeling of software evolution processes for policy investigation: Approach and example. Journal of Systems and Software 59(3), 271–281 (2001)
23. Kleijnen, J.: Sensitivity Analysis and Optimization of System Dynamics Models: Regression Analysis and Statistical Design of Experiments. System Dynamics Review 11(4), 275–288 (1995)
24. Leeflang, P., Wittink, D.: Building models for marketing decisions: Past, present and future. International Journal of Research in Marketing 17(2-3), 105–126 (2000)
25. Little, J.: Aggregate Advertising Models, The State of the Art. Operation Research 27, 629–667 (1979)
26. Mosekilde, E., Larsen, E.R., Sterman, J.D.: Coping with Complexity: Deterministic Chaos in Human Decision making Behavior. In: Casti, J.L., Karlqvist, A. (eds.) Beyond Belief: Randomness, Prediction, and Explanation in Science. CRC Press, Boston (1991)
27. Nicholson, C., Kaiser, H.: Dynamic market impacts of generic dairy advertising. Journal of Business Research 61(11), 1125–1135 (2008)
28. Otto, P.: A system dynamics model as a decision aid in evaluating and communicating complex market entry strategies. Journal of Business Research 61(11), 1173–1181 (2008)
29. Papachristos, G.: A system dynamics model of socio-technical regime transitions. Environmental Innovation and Societal Transitions 1(2), 202–233 (2011)
30. Qi, C., Chang, N.B.: System dynamics modeling for municipal water demand estimation in an urban region under uncertain economic impacts. Journal of Enviroment Management, 1628–1641 (2011)
31. Richardson, G., Otto, P.: Applications of system dynamics in marketing: Editorial. Journal of Business Research 61(11), 1099–1101 (2008)
32. Sakao, T., Lindahl, M.: Introduction to Product/Service-System Design. Springer (2010)
33. Sánchez, C., Cedillo, M.G., Villanueva, P., Martínez, J.: Global economic crisis and Mexican automotive suppliers: impacts on the labor capital. Simulation 87(8), 711–725 (2011)
34. Sheth, J., Sisodia, R.S.: Marketing productivity: issues and analysis. Journal of Business Research 55(5), 349–362 (2002)
35. Sterman, J.: Business dynamics: Systems thinking and modeling for a complex world. McGraw-Hill, Irwin (2000)
36. Suryani, E., Chou, S., Chen, C.H.: Air passenger demand forecasting and passenger terminal capacity expansion: A system dynamics framework. Expert Systems with Applications 37, 2324–2339 (2010)
37. Varey, R.: Marketing as an Interaction System. Australasian Marketing Journal 16(1), 79–94 (2008)
38. Vidale, M.L., Wolfe, H.B.: An operation research study for the response to advertising. Operation Research 5, 370–381 (1957)
39. Wierenga, B., Ophuis, P.: Marketing decision support systems: Adoption, use, and satisfaction. International Journal of Marketing Research 14(3), 275–290 (1997)
40. Yim, N., Kil, S., Kim, H., Kwahk, K.: Knowledge based decision making on higher level strategic concerns: system dynamics approach. Expert System with Applications 27(1), 143–158 (2004)

Using Caching Techniques to Improve the Performance of Rule-Based Inference Applications in Semantic Technologies

Alejandro Rodríguez-González

Centro de Biotecnología y Genómica de Plantas, Universidad Politécnica de Madrid, Spain

Abstract. Caching strategies are typical computer science tools used in diverse application fields such as the improvement of the performance of CPUs, Disks or Databases. In the field of data management, caching queries are also a widely employed technique for performance improvement that has been previously applied to XML queries. The strategy of caching specific patterns of results enables such systems to eliminate the requirement to repeat the same queries, speeding up the response time and eliminating redundancy. This chapter introduces a system designed to apply caching techniques to Semantic Technology applications. The system has been tested in a semantic diagnosis support system. Testing results are more than promising, achieving improvements in the query response time.

1 Introduction

The World Wide Web is the largest source of information. Both the information that is stored on the Web and the number of its human users have been growing exponentially in recent years and, for many people, the Web has started to play a fundamental role as a means of providing and searching for information [1]. The performance in information retrieval is a key issue in an interconnected world and many techniques have been proposed to boost this performance. Franklin argues that caching techniques play a very important role in optimizing systems performance [2]. This technique has been widely adopted in many different areas for years, therefore a great deal of effective strategies [3] and algorithms [4]. Advanced use of caching using frequent patterns recognition [5, 6] and query caching [7] have been broadly discussed in the literature. Many works focus on applying semantics to cache tables [8] or retrieving data based on already performed, different queries [9].

Given that caching techniques are extremely popular in the academic and industry field, applying caching tools to inference applications in Semantic Technologies seems to be a more innovative and defined research line. In this way, previous caching strategies have been principally applied to simple XML queries; however, the current work proposes the application of caching to queries based on logical inference.

The Semantic Web is an extension of the current Web by standards and technologies that help machines to understand the information on the Web so that they can support richer discovery, data integration, navigation, and automation of tasks [1].

T. Matsuo & R. Colomo-Palacios (Eds.): *Electronic Business and Marketing*, SCI 484, pp. 85–101.
DOI: 10.1007/978-3-642-37932-1_7   © Springer-Verlag Berlin Heidelberg 2013

Berners-Lee proposed the Semantic Web [10] as a natural evolution of the traditional Web to allow for the manipulation of content by applications with the capacity to interpret the semantics of information. Given this definition, Semantic Technologies have emerged as a new and highly promising context for knowledge and data engineering [11]. Thus, this technology promises to solve many challenging and cost-intensive problems of the current generation of Web applications [12]. The Semantic Web provides an alternative solution to represent the comprehensive meaning of integrated information and promises to lead to efficient data management by establishing a common understanding by ontologies [13].

The power of publishing and linking data in a way that machines can automatically interpret through ontologies is beginning to materialize [13], and as semantic web achieve a feasible level of maturity, they become increasingly accepted in various business settings at enterprise level [12]. Moreover, according to Lytras and García, in recent years, Semantic Web research has resulted in significant outcomes and the adoption of this technology from the market and the industry is becoming closer [14]. In other words, semantic web is beginning to be pragmatic [15].

Due to this importance, performance issues about semantic web applications are beginning to arise (e.g. [16, 17]), and this is the trend that this chapter follows.

2   Background

Both if the Web is still static content-based (based on HTML and PDF) or semantically enriched and dynamic, the Web systems performance is very relevant and might become a threat. In the semantic web scenario, the performance of highly globalized applications is also a big issue for both researchers and developers. In response to this necessity several recent works deal with performance in Semantic Web (e.g. [16, 18, 19]). One of the solutions given to performance threat is caching those contents that are more demanded using pattern recognition in both contents and queries.

In the following, several works about query pattern recognition both in semantic and non semantic environments are depicted.

Many works have dealt with the discovery of patterns in XML Queries (e.g. [6, 20]), however, given the interconnected and interlinked nature of semantic content, these efforts are far from being applicable in semantic web scenario. In the work of Kim, an inference-based web ontology that uses expression of Description Logic and Semantic Web Rule Language (SWRL) is presented [21]. This ontology was developed in order to build an Intelligent Image Retrieval System. A new approach for ontologies mapping and inference techniques of description logic are proposed in [22]. The ontologies mapping from the computing of linguistic or structural similarities to the logical reasoning between concepts of different ontologies are transformed, implementing the semantic mapping of ontologies. This approach is based on the CTXMATCH algorithm (propositional logic based algorithm) proposed by Bouquet et al. [23]. However, given the ontology, which defines diverse relations and restrictions among concepts, the power of description and reasoning of propositional logic is not enough for general purposes. In [24], the authors presented a prototype that is capable of drawing plausible inferences from Resource Description Framework

(RDF), which are constituents of a distributed, Semantic Web knowledge system. The approach taken to build the prototype seems to be the extension and adaptation of a classical approach to plausible inference to exploit the evolving infrastructure being developed to represent declarative knowledge on the Semantic Web. The approach includes a knowledge representation formalism that supports meta properties, which define precise semantics, enabling subsequent plausible inferences via extended composition of RDF properties. In [25], a prototype known as WebSys is presented. This prototype provides resource information service for resource discovery, resource matching and resource monitoring, in order to allocate appropriate resource for user task. WebSys allows users to build a manifold and reliable information processing service by enabling the incorporation of various semantic ontologies for the inference of semantic information.

A system to facilitate exchange of information by automatically finding experts, competent in answering a given question is proposed by [26]. The advantage of the proposed system over standard forums or group-ware systems is that full-formatted questions can be compared to stored qualification profiles, which were automatically derived from documents, without Human search effort, and possibly refined manually. This allows us to find competent colleagues (or helpful literature such as How-Tos) for a given problem in a single step, and without intermediate iterations. The system is symmetric in that it does not distinguish between "questioners" (asking questions) and "experts" (answering them), therefore forming a "community" of users, which are distributed over an ontology covering the total knowledge. This ontology can either be given (i.e. in the form of an organizational chart), or it can be derived from the experts' knowledge.

3   Main Focus of the Chapter

Nowadays there are a huge variety of rule based systems that are used as expert systems, decision support systems, etc. The vast majority of these systems don't focus in the problem that represents the increase of the number of rules that they should manage. In computational terms, the increase of the size of the knowledge base where the rules are stored in (that can be achieved by the increase in the number of rules or the increase in the extension of the rules already stored), is an important fact that should be taken into account in the development of rule based systems.

However, the resolution of this problem not always is a trivial task. Depending the kind of system developed, the domain, and some other parameters, it is not an easy task the aim of reducing the size of rules knowledge base (reduce the number of rules, or reduce the extension or length of them).

For this reason, in this work, we describe a concrete system that works over a concrete domain (which can be extrapolated to many other different domains), that allows to propose a solution to the problem addressed.

The idea of this system is generate an architecture that is able to find patterns in the queries that are made against the system, and with these patterns, generate new subsets of rules. Once the new subset of rules are generated, the system will be able to choose a reduced set of rules, in order to make inference, ensuring that the provided results will be correct and reducing time.

3.1   Pattern Caching Strategy (PCS)

The pattern-based caching strategy consists of searching within the table of cached elements, with the objective of extract concrete patterns from the input data. This process is carried out systematically by an agent who is responsible for the search and retrieval of these patterns. Once one or several patterns have been found, the retrieved information is processed, with the aim of establishing whether the patterns are valid or adequate enough to enable an increase in the system efficiency.

The objective of this strategy is to notably increase the efficiency of the system. The first step is an inference process which is supported by a knowledge base represented with ontology, and another base which consist in a set of rules, both related by means of the inference engine. This inference process generally has a determined non-linear time or cost, which increases as the size of the knowledge base increases, as well as the knowledge base which represents the rules (the number of rules).

Once the PCS has been applied, a number of patterns of inputs which are used with a given frequency can be found. Therefore, the objective is to generate new sets of rules based on these patterns, in such a way that when the inference engine receives a new query as input, the engine verifies whether this query matches any previously found patterns prior to performing the inference process. In the case that a match is found between the input and the stored patterns, the system assigns this query to the corresponding knowledge base of rules, considerably increasing the performance of the system given that it is not required to verify the input querying the global set of rules, since it is of greater size and whose inference time is much longer.

3.2   Preconditions of the System

Some basic preconditions for the efficient functioning of this type of caching strategies should be outlined. The principal and most important feature is the following: based on the output of the inference system, it can be possible to obtain new inference rules which generate the same results. This characteristic is required, because once an inference pattern has been discovered, all of the possible results with this pattern need to be determined. Based on these results, a new set of rules is generated which is smaller than the initial set. These rules lead to the outcomes which have generated the result just obtained. In principle, that could be summarized through the following formula:

Set of rules for the pattern $\subset$ *Set of total rules*

The set established above is always smaller than the total set. It is precisely this characteristic which enables an increased velocity and efficiency of the system, when the inference engine is executed with a smaller and more manageable set of rules.

Therefore, it is essential to establish that the outputs of the system enables the generation of inference rules, that is, they allow the determination of the premises which lead to the conclusion. Another issue which should be considered is that this solution is only applicable to systems in which the number of inputs directly affects the number of outputs. More specifically, if the system receives X inputs and returns

Y outputs, in the case that it receives these X inputs a second time with an additional input element, the number of outputs in this case should be less than Y, and additionally, should be elements of Y. This is due to the fact that the new inputs which have been added to the set X have redefined the result such that it is a more closed set than the previous set, as it is shown in the following:

$$X \ (inputs) \rightarrow infers \rightarrow Y \ (outputs)$$
$$X + T \ (inputs) \rightarrow infers \rightarrow Z \ (outputs)$$
$$Where: Z \subset Y$$

The rule language used was Jena Rules [27], but this system can be applied for any kind of rule language like SWRL or others. The inference engine also is the Jena rule engine.

3.3 Global Components of the System

The system has been developed to be used in applications which receive many queries to their inference engine, which is reflected in the global performance of the system, noticeably slowing it down and degrading its functioning due to the reasoning or inference process. Thus, the system has been developed as shown in the following figure:

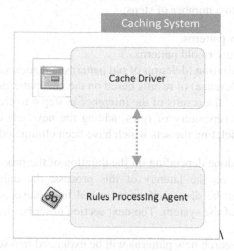

Fig. 1. Caching System

In the figure 1, two fundamental constituents can be viewed in the Caching System component.

3.4 Cache Driver

The Cache Driver is responsible for managing the cache and all of its related objects (it's Database, content, etc.) It ensures that the number of elements is adequate, and manages the replacement of the elements (Most Recent Used (MRU), Least Recent Used (LRU), etc.).

Its basic functioning is that of a mere container, which interacts with the rest of the systems, such as the processing of the rules or the inference engine.

The main working is based on its interaction with the inference engine. The engine is in charge for verifying with the Cache Driver if the input has a pattern in the cache that can match with it. In the case that they are inside the cache, this indicates that a search should be performed for a subset of rules which is not the global set, therefore functioning to obtain the knowledge in a more efficient time span.

Furthermore, this component is in charge of verifying that the size of the cache does not exceed a concrete threshold, deleting tuples based on specific criteria. Generally, the criteria to be taken into account are that the tuples which contain more "hits" are those more likely to survive. The tuples which are consulted with a greater frequency are characterized by a greater number of hits.

3.5   Rules Processing Agent

The rules processing agent synchronizes the global repository of rules (see Fig. 3, architecture) based on the data which it receives from the Cache Driver. Its fundamental aim is to get the patterns of inputs which are stored in the cache and create new sets of rules, which will be synchronized according to the global repository. This process is divided into a number of steps:

1.   Search for new patterns.
2.   Update of the use of old patterns.
3.   Cache reorganization (deletion of old patterns and addition of new ones).
4.   Reasoning (inference) of results based on the new patterns.
5.   Computation of the results of the inference in step 4 to create new sets of rules.
6.   Update of the repository of rules, adding the new sets created based on new patterns, and deleting the sets which have been eliminated.

This process can be done depending on the duration of the process over every X units of time. Depending on the latency of this process, the units of time change so that parameter X can be decreased or increased to achieve an improvement in the global performance of the system. The next section describes the algorithm of pattern recognition.

The algorithm to search new patterns will be explained in a set of steps:

1. First of all, get all the hashes of BD. Imagine these 6 hashes:

55f5a7216d1c107c62508bc690d78392
e4673bd7788f8a1b55266f3f77464b1f
444096cb35814b0857df5fbf6d3372d2
335c8c67e7bda3abe46b5a0dbd645bbd
67323fa32b004ea1013dbf444fe976cf
9c975b4518210e6fecb874559e3027ec

2. For every hash the system will generate an object (Caching Consult) that encapsulates the hash and their corresponding codes associated. In our example (medical system), will be symptoms, so we will have:

55f5a7216d1c107c62508bc690d78392 - *sA*
e4673bd7788f8a1b55266f3f77464b1f - *sA,sB,sC*
444096cb35814b0857df5fbf6d3372d2 - *sA,sB,sD*
335c8c67e7bda3abe46b5a0dbd645bbd - *sA,sB,sD,sF*
67323fa32b004ea1013dbf444fe976cf - *sG,sH,sI*
9c975b4518210e6fecb874559e3027ec - *sG,sH,sI,sL*

In this case, we are able to identify some patterns. The pattern depends of the length that we set that should have it. In this case, we establish at least 3 elements to build a pattern.

3. Now, we create a list of strings that contains only the codes of the previous objects. So, we will have:

sA
sA,sB,sC
sA,sB,sD
sA,sB,sD,sF
sG,sH,sI
sG,sH,sI,sL

4. Now, we will take every previous string and we compare each one with the rest (with some restrictions), trying to find matches. So, the trace of this part of the al gorithm will be:

sA,sB,sC / sA,sB,sD -> (Match: 2 / Discard. Minimum: 3)
sA,sB,sC / sA,sB,sD,sF -> (Match: 2 / Discard. Minimum: 3)
sA,sB,sC / sG,sH,sI -> (Match: 0 / Discard. Minimum: 3)
sA,sB,sC / sG,sH,sI,sL -> (Match: 0 / Discard. Minimum: 3)

sA,sB,sD / sA,sB,sD,sF -> (Match: 3 / Pattern found: sA,sB,sD)
sA,sB,sD / sG,sH,sI -> (Match: 0 / Discard. Minimum: 3)
sA,sB,sD / sG,sH,sI,sL -> (Match: 0 / Discard. Minimum: 3)

sA,sB,sD,sF / sG,sH,sI -> (Match: 0 / Discard. Minimum: 3)
sA,sB,sD,sF / sG,sH,sI,sL -> (Match: 0 / Discard. Minimum: 3)

sG,sH,sI / sG,sH,sI,sL -> (Match: 3 / Pattern found: sG,sH,sI)

The algorithm discard the elements that didn't have enough elements to be a pattern (for example "sA" that is a single element and algorithm establish that are needed at least three elements). Also, it also doesn't check elements that have already checked. For example: (sA,sB,sC / sA,sB,sD is the same that check sA,sB,sD / sA,sB,sC).

3.6   Pattern Recognition Algorithm

One of the main components of the system is the determination of patterns or sequences of data which can be repeated in the searches which users are likely to perform. In order to achieve this functionality, an algorithm should be applied which is capable of traversing the queries stored in the cache and storing the patterns which are repeated, to subsequently utilize this information in a way which considerably optimizes search time.

The algorithm which enables the localizing of these patterns should browse every single sequence of input data, so that necessary connections can be done between them, and enable the location and storage of the patterns which will subsequently be searched for.

There are various forms of realizing this task, one of which is the following. In the first place, the maximum length of each pattern should be established. Once this value has been fixed, the first input to the cache should be processed, and a combination of all of the elements of the set of inputs should be made, without redundancy. This combination of input data, which has an established length, forms the pattern. This pattern will be stored and compared with the future entries of the table.

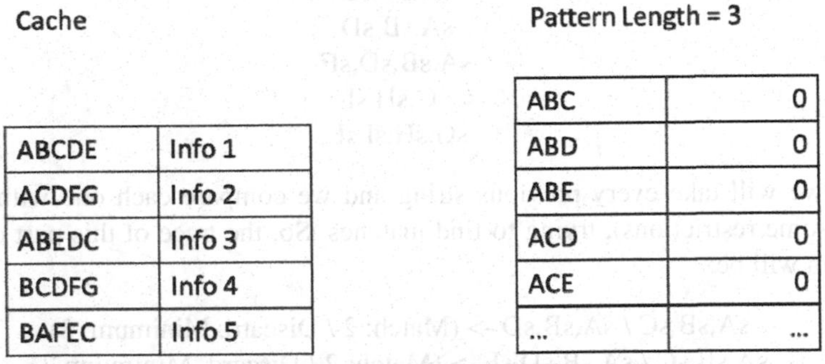

Fig. 2. Cache patterns

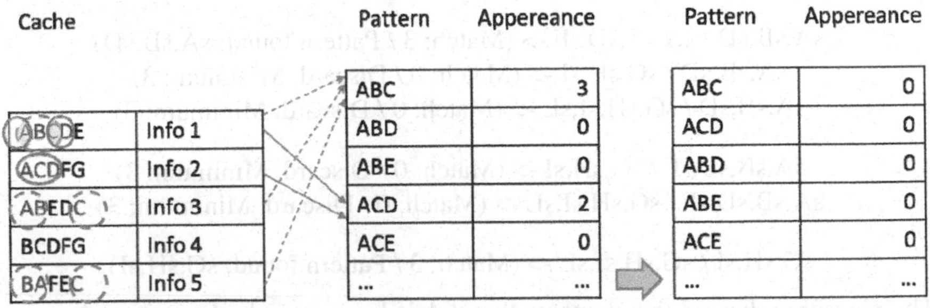

Fig. 3. Pattern recognition

Each time a pattern is found, its occurrence will be added to the table, increasing in its ranking, producing the result that each time more queries are made, the most frequent queries are stored, with a higher ranking. Once the entire table has been browsed and all possible patterns have been found, these are stored separately, such that in future executions of the system, comparisons are only made with these established patterns, saving the time which would be used if the entire dataset in the table was processed again.

Another form of confronting the problem of patterns location is ordering alphabetically all the input data introduced in each tuple. Once this has been carried out, the system searches for patterns which match with the existing ones. The advantage that this system offers is that it is possible to eliminate unnecessary accesses to the system, given that when positions are systematically compared, if one of them does not match, it is not necessary to continue processing, as occurs with an algorithm which does not entail caching.

However, even given this scenario, all of the algorithms which aim to solve these types of problems are rather slow, given that they should transverse all of the tuples in the table. The type of data managed in the current case aids the fact that the process does not slow down excessively, and allows the management of cost at an acceptable level.

With regard to related algorithms, various exist whose function is to search for chains. In particular, many of these types of algorithms exist in the biology domain, and are particularly related to searching for patterns in DNA chains [28]. Occurrences of such algorithms also exist in other areas, such as those which analyze the traffic in a network, and attempt to extract behavioral patterns to be able to develop corresponding response strategies [29].

3.7   System Architecture

The architecture of the proposed system is a traditional Expert System: a Knowledge Base, a set of inference rules and an inference engine. The novel feature is the inclusion of a caching system for the creation of new subsets of rules, in the form of minor modifications in the rules container, which is composed of two repositories, one global which stores all of the initial rules of the system, and another which is responsible for storing the distinct subsets of rules which will be generated by the Caching System.

The figure which is shown below shows the logical architecture of the system, where it can be seen how the different subsystems are communicated, as well as the flow of messages which are interchanged in order to realize the inference process. In the next sections, the internal functioning of each of these elements will be described, paying special attention to the Rules Manager and Caching System, the components which represent the core of this work.

3.8   Knowledge Base

This component represents the Knowledge Base [30] of the Expert System which is used to carry out the reasoning process. Concretely, it is the ontology manager [31] of the system. It provides all of the data that it has stored in the repository, in such way that the inference engine can perform with explicit knowledge about the used domain.

Thus, it receives new facts from the engine, which are the result of constant interaction and reasoning which this base orchestrates.

3.9   Inference Engine

The inference engine is the core of the Expert System [32], which carries out the entire reasoning and inference process to generate an answer based on the experience and knowledge that it contains.

At the same time, it has the aim of generating the hash codes for each input which it receives; with the objective that they will be stored in the cache and will be consulted later if it is possible to obtain a subset of rules which is capable of solve the problem.

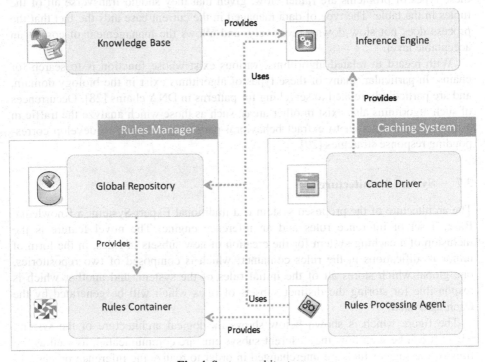

Fig. 4. System architecture

3.10   Rules Manager

It transfers the rules which provide the restrictions for the inference process to the inference engine, working over the Knowledge Base. It is divided into two subsystems, the first one being the global repository, which is comprised of the total set of rules of the system, while the Rules Container systematically stores the distinct subsets of rules that are generated through the caching system.

3.11   Global Repository

This repository stores the global inference rules, being the repository of the rules which affect the entire problem domain, and the one that will be increased as new facts are introduced or as the application domain of the system increases. It is managed by the expert, and can only be modified if the expert extends or reduces the magnitude of the problem domain.

3.12   Rules Container

It is a repository where the distinct subsets of rules are stored, generated by interaction with the Caching System module. The Agent, based on the Caching System, analyses the total set of rules, searching for patterns which are repeated in the cache table, and which are determined based on a concrete group of rules. In this way, the rules which are not necessary for this step are deleted, reducing in a considerably way the size of the problem domain, thus decreasing the response time of the system.

3.13   Caching System

As outlined in Global components of the system, the goal of this module is to reduce the size of the problem, decreasing the number of rules which the inference engine has to process, and notably limiting the magnitude of the domain in question.

3.14   Cache Driver

The Cache Driver is the subsystem in charge of the management of the cache results. It provides an access interface to the inference engine which it enables it to consult in this database whether a pattern exists which corresponds to the current query. As a consequence, it allows access to the Rules Container to obtain a limited subset over which it can work.

3.15   Rules Processing Agent

This module is in charge of carry out the processing of the global set of rules to generate new subsets which can solve the problem based on the entries stored in the cache, as well as the management of their storage. It is implemented by means of an Intelligent Autonomous Agent [33, 34], which, at systematic moments in time, carries out a map of the cache in search of new patterns, freeing space within it in the case that it is required. The method selected for this process is LFU (Least Frequently Used) [35], given that it abolishes the least usual combinations and those that have a lower probability of occurrence. As a consequence, they have a higher or lower impact on the optimization of the system. The agent will be responsible for searching for combinations within the Global Repository which can match the pattern in question, storing these new resulting subsets in the Rules Container [36]. At the same time, in the case of carrying out deletions in the cache table, it will also have as a goal to abolish those subsets which correspond to the deleted patterns.

4   Use Case

Current section describes an example of system working. In this problem, we use a medical diagnosis knowledge base as domain [37]. As was mentioned previously, in the case where rules container module contains a high number of rules, it is highly possible that the performance or speed of the system will be affected. For this reason, the advantage of the use of this system is based in the use of be able to work with a reduced set of rules, which will imply that the performance of the system will be increased.

The inputs that system receives in this use example are the codes of the symptoms that normally the system receives to make a diagnosis. Each input is uniquely represented by a single string. That strings, will be separated by comas. For simplification purposes, the outputs will be also strings.

Bearing the example in mind, the table below represents an example of the cache in the current system:

Note: SYM_A = sA, SYM_B = sB, etc. (To clarify with previous notation).

Table 1. Use case. Cache table

Hash	Input	Output	Times	Date
2b0dc568e588	SYM_A	*DIS_M,DIS_N,DIS_P,DIS_T,DIS_X,DIS_Z*	356	10/01/08
e86410fa2d6e	SYM_A,SYM_B,SYM_D,SYM_F	*DIS_M*	245	10/01/08
8827a41122a5	SYM_G,SYM_H,SYM_I,SYM_L	*DIS_W*	240	09/01/08
ab5d511f23bd	SYM_A,SYM_B,SYM_C,SYM_E	*DIS_N*	145	09/01/08
3c6f5dee2378	**SYM_A,SYM_B, SYM_C**	*DIS_N,DIS_X*	98	08/01/08
6b642164604d	**SYM_G,SYM_H, SYM_I**	*DIS_V,DIS_W*	98	08/01/08
8c328f9a099b6	**SYM_A,SYM_B, SYM_D**	*DIS_M,DIS_Z*	98	08/01/08

The above table represents an example of the approximated content of an individual caching strategy. Notice that a number of patterns has been found in the inputs and are indicated in bold.

The system will carry out the strategy previously described with this set of patterns. In this case, the system will search possible common patterns within all of the inputs, and it would find the patterns as displayed in the current example. Once the system has extracted all of the new patterns (if a pattern is extracted which is already present in the system, it will be omitted, as it already exists), the system will communicate with the inference engine to attempt to obtain all of the possible outputs which it (or the pattern) would provide. In the current example, the system will make a query to the inference engine so that it returns all of the outputs based on the set of inputs of {SYM_A, SYM_B, SYM_C}, {SYM_G, SYM_H, SYM_I} and {SYM_A, SYM_B, SYM_D}.

The system, based on this inputs, returns the following:

$$SYM\_A, SYM\_B, SYM\_C \rightarrow DIS\_X, DIS\_N$$
$$SYM\_G, SYM\_H, SYM\_I \rightarrow DIS\_V, DIS\_W$$
$$SYM\_A, SYM\_B, SYM\_D \rightarrow DIS\_Z, DIS\_M$$

Looking at the outputs that the system has returned, it can be observed that these are also present in all of the elements that have been output of the inference engine. It is evident that more elements exist than those that are in the table, but it is important that all of the elements that are displayed in the table are present. This situation is the result of one of the preconditions discussed previously.

Once this step has been performed, the goal of the system is to generate all of the possible inference rules that lead to this output based on the output that it has just obtained.

5   Evaluation

Evaluation has been done taking making some test and measuring the times of inference. In order to contrast the results in a better way, two kind of tests where developed. The first one measure the execution time of the entire system that we are using [37], taking two time references: time using the caching system and time without using it. The second one, only takes into account the inference time (leaving behind rest of system like data loading or others), because is the real time that we want to improve.

For both kind of test, two kinds of executions were tested. First one is executing the test (system or inference engine) once. Second one is executing the test 100 times.

For the evaluation, three data sets were used. The data sets created follows the guidelines used in the use example:

- DS1 (Data set 1): The first data set contains the codes of symptoms "A, B, D, F" with the intention of use subset created with the pattern "A, B, D".
- DS2 (Data set 2): The second data set contains the codes of symptoms "G, H, I, L" with the intention of use subset created with the pattern "G, H, I".
- DS3 (Data set 3): The third data set contains the codes of symptoms "A, B, C, E" with the intention of use subset created with the pattern "A, B, C".

Next tables show the results of the execution of the entire system.

Table 2. System test. One execution

Data Set	Time Caching Disabled	Time Caching Enabled
DS 1	219 ms	234 ms
DS 2	188 ms	188 ms
DS 3	156 ms	156 ms

Table 3. System test. One hundred executions

Data Set	Time Caching Disabled	Time Caching Enabled
DS 1	7 ms	6 ms
DS 2	4 ms	4 ms
DS 3	5 ms	5 ms

Table 2 shows the execution of the entire system for the very first time. As it can be seen in the table and the associated graphic, one single execution does not improve inference times, and even in one case, the caching system decreases the performance of the system. The reason of this is because in one execution, the inference system takes more time to load all the data and to compare it against the ontology. If we also add the time of the pattern search algorithm which try to match the given input with a pattern of the caching we get as result the increase of the times in one single execution.

However, in table 3 can be seen how the times in 100 executions shows an increase in the efficiency in the system. For example, the case of test with data set 1 (DS 1) shows a time of 6 ms in the execution of the system with the caching enabled. That means, that 100 executions take a time of 600 ms (6ms x 100), giving a time for a single execution of 6ms. Compared with the first case, in one single execution, the difference is about 238 ms (234ms - 6ms).

Furthermore, it should be taken into account that this is the time for the entire system, and not for the inference engine, which is the system whose performance we are trying to improve.

Next tables show the results of the execution of the inference engine.

Table 4. Inference test. One execution

Data Set	Time Caching Disabled	Time Caching Enabled
DS 1	16 ms	15 ms
DS 2	32 ms	16 ms
DS 3	0.9 ms	0.7 ms

In this comparative we can see the real performance of the caching system. In the first case, we can see that in two cases the times are improved. In one of them, with a reduction of one millisecond of inference time, and in other, with a reduction of 50% (16ms).

Table 5. Inference test. One hundred executions

Data Set	Time Caching Disabled	Time Caching Enabled
DS 1	1 ms	0.7 ms
DS 2	0.8 ms	0.6 ms
DS 3	0.9 ms	0.5 ms

With this data, we can conclude that the caching system gets an increment of temporal efficiency for small data sets. This evaluation has been done with this type of sets because the system where the caching system was integrated and tested it's hard to introduce bigger datasets in an automatic way to make an evaluation with it.

6   Future Research Directions

Applications where their main internal working is a rule based system normally have a lack in efficiency time. Depending on the size of the knowledge base the time of a consult can vary from a few seconds to a several minutes. The reason of this enormous time variation are directly related with the size of the knowledge base, and for this reason it is neccesary to generate system that allows to reduce the concepts stored in the knowledge base of rule based systems, or find algorithms that allows to make inference with a reduction of reasoning time.

Future work will include the challenging task of optimizing the pattern search algorithm. However, it is not an easy task because, as stated previously, in every case it will be necessary to traverse the entire set of tuples that comprise the cache.

7   Conclusion

The generalization of the use of Semantic Web application in industrial environments strongly requests new data-intensive management tools to manage the tremendous collections of data that such applications handle. Thus, the size of the Web, and consequently the size of the Semantic Web is expected to grow dramatically in the coming years and performance of these systems becomes increasingly important. As a result of this fact, many recent works are devoted to semantic web performance (e.g. [18, 19]). However, most of current semantic data management solutions present poor performance and low scalability since most queries require multiple self-joins on the triples table [16]. To avoid that the solution to information overload constitutes a problem in performance and efficiency, this chapter introduced the use of caching techniques in the management of semantic data. Following the path of previous works about efficient mining of query patterns for caching purposes, it is proposed a Pattern Caching Strategy and a Pattern Recognition Algorithm, and once query patterns are detected, caching can be effectively adopted for query performance enhancement. Experimental results demonstrate that the solution presented in this chapter can achieve high performance for query evaluation, and it is efficient and scalable, and caching the results of frequent patterns significantly improves the query response time.

References

1. Eiter, T., Ianni, G., Lukasiewicz, T., Schindlauer, R., Tompits, H.: Combining answer set programming with description logics for the Semantic Web. Artificial Intelligence 172, 1495–1539 (2008)
2. Franklin, M.J.: Client Data Caching: A Foundation for High Performance Object Database Systems. Springer (2011)
3. Alonso, R., Barbara, D., Garcia-Molina, H.: Data caching issues in an information retrieval system. ACM Trans. Database Syst. 15, 359–384 (1990)
4. Johnson, T., Shasha, D.: 2Q: a low overhead high performance buffer management replacement algorithm. Presented at the Proceedings of the Twentieth International Conference on Very Large Databases (1994)
5. Bei, Y., Chen, G., Hu, T., Dong, J.: A Caching System for XML Queries Using Frequent Query Patterns. In: 11th International Conference on Computer Supported Cooperative Work in Design, CSCWD 2007, pp. 47–52 (2007)
6. Yang, L.H., Lee, M.L., Hsu, W., Huang, D., Wong, L.: Efficient mining of frequent XML query patterns with repeating-siblings. Information and Software Technology 50, 375–389 (2008)
7. Adali, S., Candan, K.S., Papakonstantinou, Y., Subrahmanian, V.S.: Query Caching and Optimization in Distributed Mediator Systems. In: Proc. of ACM SIGMOD Conf. on Management of Data, pp. 137–148 (1996)
8. Ren, Q., Dunham, M.H., Kumar, V.: Semantic Caching and Query Processing. IEEE Trans. on Knowl. and Data Eng. 15, 192–210 (2003)
9. Godfrey, P., Gryz, J.: Semantic Query Caching for Heterogeneous Databases. In: Proceedings KRDB at VLDB 1997, pp. 6–1 (1997)
10. Berners-Lee, T., Hendler, J., Lassila, O.: The Semantic Web. Scientific American 284(5), 34–44 (2001)
11. Vossen, G., Lytras, M., Koudas, N.: Editorial: Revisiting the (Machine) Semantic Web: The Missing Layers for the Human Semantic Web. IEEE Transactions on Knowledge and Data Engineering 19, 145–148 (2007)
12. Nixon, L.J.B., Simperl, E., Krummenacher, R., Martin-recuerda, F.: Tuplespace-based computing for the semantic web: A survey of the state-of-the-art. Knowl. Eng. Rev. 23, 181–212 (2008)
13. Shadbolt, N., Berners-Lee, T., Hall, W.: The Semantic Web Revisited. IEEE Intelligent Systems 21, 96–101 (2006)
14. Lytras, M.D., Garcia, R.: Semantic Web applications: a framework for industry and business exploitation – What is needed for the adoption of the Semantic Web from the market and industry. International Journal of Knowledge and Learning 4, 93–108 (2008)
15. Alani, H., Hall, W., O'Hara, K., Shadbolt, N., Szomszor, M., Chandler, P.: Building a Pragmatic Semantic Web. IEEE Intelligent Systems 23, 61–68 (2008)
16. Abadi, D.J., Marcus, A., Madden, S.R., Hollenbach, K.: SW-Store: a vertically partitioned DBMS for Semantic Web data management. The VLDB Journal 18, 385–406 (2009)
17. Wu, G., Li, J., Hu, J., Wang, K.: System II: A Native RDF Repository Based on the Hypergraph Representation for RDF Data Model. Presented at the July (2008)
18. Hellmann, S., Lehmann, J., Auer, S.: Learning of OWL Class Descriptions on Very Large Knowledge Bases. International Journal on Semantic Web and Information Systems 5, 25–48 (2009)
19. Hogan, A., Harth, A., Polleres, A.: Scalable Authoritative OWL Reasoning for the Web (2009)

20. Theodoratos, D., Wu, X.: Assigning semantics to partial tree-pattern queries. Data Knowl. Eng. 64, 242–265 (2008)
21. Kim, S.-K.: Implementation of Web Ontology for Semantic Web Application. In: Proceedings of the Sixth International Conference on Advanced Language Processing and Web Information Technology (ALPIT 2007), pp. 159–164. IEEE Computer Society, Washington, DC (2007)
22. Ma, W., Xu, G., Wang, G., Liu, J.: Detecting Semantic Mapping of Ontologies with Inference of Description Logic. Presented at the (2008)
23. Bouquet, P., Serafini, L., Zanobini, S.: Semantic coordination: A new approach and an application. In: Fensel, D., Sycara, K., Mylopoulos, J. (eds.) ISWC 2003. LNCS, vol. 2870, pp. 130–145. Springer, Heidelberg (2003)
24. Shrivastava, S., Goudar, R.H., Aital, P.: A Plausible Inference Applied to the Mechanism of Semantic Web Searching. In: Proceedings of the 2008 First International Conference on Emerging Trends in Engineering and Technology, pp. 1136–1139. IEEE Computer Society, Washington, DC (2008)
25. Kim, T.-N., Kim, H.-L., Yi, K.-H., Jeong, C.-S.: WebSIS: Semantic Information System Based on Web Service and Ontology for Grid Computing Environment. Presented at the (October 2007)
26. Metze, F., Bauckhage, C., Alpcan, T.: The "Spree" Expert Finding System. In: Proceedings of the International Conference on Semantic Computing, pp. 551–558. IEEE Computer Society, Washington, DC (2007)
27. Dave Reynolds: JUC – Jena Rules. Hewlett-Packard Development Company (2004)
28. Smith, T.F., Waterman, M.S.: Identification of common molecular subsequences. J. Mol. Biol. 147, 195–197 (1981)
29. May, P., Ehrlich, H.-C., Steinke, T.: ZIB structure prediction pipeline: Composing a complex biological workflow through web services. In: Nagel, W.E., Walter, W.V., Lehner, W. (eds.) Euro-Par 2006. LNCS, vol. 4128, pp. 1148–1158. Springer, Heidelberg (2006)
30. Gruber, T.R.: Toward Principles for the Design of Ontologies used for Knowledge Sharing. International Journal of Human-Computer Studies 43, 907–928 (1995)
31. Guarino, N.: Formal Ontology in Information Systems. In: Proceedings of the 1st International Conference, Trento, Italy, June 6-8. IOS Press, Amsterdam (1998)
32. Hayes-Roth, F., Waterman, D.A., Lenat, D.B.: Building expert systems. Addison-Wesley Longman Publishing Co., Inc., Boston (1983)
33. Luke, S., Spector, L., Rager, D., Hendler, J.: Ontology-based Web agents. In: Proceedings of the First International Conference on Autonomous Agents, pp. 59–66. ACM, New York (1997)
34. Girardi, R., Faria, C.G.D., Balby, R.: Ontology-based Domain Modeling of Multi-Agent Systems
35. Chang, R.-S., Chang, H.-P., Wang, Y.-T.: A dynamic weighted data replication strategy in data grids. In: IEEE/ACS International Conference on Computer Systems and Applications, AICCSA 2008, pp. 414–421 (2008)
36. Wang, H., Kwong, S., Jin, Y., Wei, W., Man, K.-F.: Agent-based evolutionary approach for interpretable rule-based knowledge extraction. IEEE Transactions on Systems, Man, and Cybernetics, Part C: Applications and Reviews 35, 143–155 (2005)
37. García-Crespo, Á., Rodríguez, A., Mencke, M., Gómez-Berbís, J.M., Colomo-Palacios, R.: ODDIN: Ontology-driven differential diagnosis based on logical inference and probabilistic refinements. Expert Systems with Applications 37, 2621–2628 (2010)

Influence of Users' Perceived Compatibility and Their Prior Experience on B2C e-Commerce Acceptance

Ángel Herrero Crespo, Mª Mar García de los Salmones Sánchez,
and Ignacio Rodríguez del Bosque

Department of Business Administration, Faculty of Economics, University of Cantabria,
Avda. de los Castros, s/n, 39005, Santander, Spain
{Herreroa,gsalmonm,rbosque}@unican.es

Abstract. This paper analyses the factors that determine the adoption of electronic commerce by consumers, examining the influence of two variables that have rarely been studied in the field of online purchasing: users' perceived compatibility with e-commerce and their prior experience of online purchasing. With this aim, we consider the Technology Acceptance Model –TAM– [1,2] as a reference framework. The research sample is divided between those Internet users that have never purchased on the Internet and those that have already purchased online in the past. The results obtained support that perceived compatibility has a positive influence on attitudes towards e-commerce and on perceived usefulness and on ease of use of online purchasing for Internet users with and without prior experience with online transactions. Additionally, the results obtained show off that experience with e-commerce has a moderator effect both on the causal relationships included in the TAM and the influence of perceived compatibility.

Keywords: Electronic commerce, perceived compatibility, previous experience, Internet purchasing intention, Technology Acceptance Model.

1 Introduction

The Internet provides companies with new mechanisms to access, manage and communicate information within the market, giving place to new relationships between suppliers and consumers. This technological revolution has propitiated a new business paradigm, novel business models and generating optimistic expectations for e-commerce. Nevertheless, these promising forecasts have not been accomplished and the expansion of electronic purchasing is being slower than originally expected. Therefore, the complexity and dynamism of Internet and e-commerce make evident the need to continue with the research on this topic, revising the traditional paradigm of Marketing, and studying the behavior of firms and consumers in electronic environments [3-5].

In this context, this paper intends to analyse the factors that determine the adoption of electronic commerce by consumers and which lead Internet surfers to become online purchasers. In particular, we examine in depth the influence of two variables that

T. Matsuo & R. Colomo-Palacios (Eds.): *Electronic Business and Marketing*, SCI 484, pp. 103–123.
DOI: 10.1007/978-3-642-37932-1_8   © Springer-Verlag Berlin Heidelberg 2013

have been identified as determinants of online purchasing behavior: compatibility perceived in e-commerce and prior experience of e-purchasing.

Perceived compatibility with previous values and habits has been identified within the literature about information systems (IS) as one of the main attributes of a new technology or application that determine its acceptance by users [6-12]. Nevertheless, the literature about the Internet and e-commerce has paid little attention to this variable [13]. In this sense, different studies have considered determinants of e-commerce which are conceptually related to perceived compatibility, such as prior home-shopping habits [14-17], prior knowledge of the Internet [16,18-21], a "wired lifestyle" [22] or prior use of the Web as a communication medium [14,23-27]. However, few studies have explicitly examined the influence of perceived compatibility on online purchasing behavior [13]. Additionally, these studies have adopted diverse theoretical perspectives, considering different dependant variables. Therefore, the empirical evidence available regarding the influence of perceived compatibility on online purchasing behavior is both scarce and inconclusive.

Additionally, the classical theory of consumer behavior [28,29] and attitude formation [30,31] supports the influence that individuals' experience with a conduct or product exerts on their future behavior. Accordingly, diverse authors have observed that the variables that influence the consumers' decision to make a purchase through the Internet —or the intensity of the effect exerted by these variables— may be different depending on their experience of e-commerce [13,32-36]. In particular, there may be relevant differences in the first decision to purchase a product or service online with regard to later repetitions of this behavior. Thus, both decisions are distinct in a fundamental factor: the source of the information that generates subjects' beliefs and attitudes, which, in the end, give place to a behavior or purchase decision. Thus, while the first purchase on the Internet is supported on indirect information regarding the system, the reiteration of the behavior is influenced by the individuals' experience with previous transactions on the Internet. Therefore, in this article we intend to examine how prior experience with online purchasing mediates the influence exerted by other explanatory variables (such as attitudes and beliefs) on Internet purchase behavior.

To analyse the influence exerted by users' perceived compatibility and prior experience of online purchasing, the Technology Acceptance Model (TAM) [1,2] is taken as a reference framework. The selection of this theoretical basis is justified by its extensive use in the literature about information systems acceptance and, particularly, in the field of the Internet (see [37]). Therefore, we propose an extended model of online purchase, adding prior experience of e-commerce and perceived compatibility with previous values and habits, to the traditional variables included in the TAM —attitude, perceived usefulness and perceived ease of use—. Accordingly, no hypotheses are enounced to test the widely contrasted causal relationships in TAM. However, the variables included in TAM —beliefs about technology, attitude and acceptance intention— are considered as dependant variables to contrast the influence exerted by perceived compatibility and previous experience with e-commerce.

Therefore, this paper provides three main contributions with regard to existing literature: 1) to examine the influence that perceived compatibility with previous values and habits exerts on future intention to purchase on the Internet; 2) to analyse the moderating influence exerted by previous experience with e-commerce on the effect of attitudes and beliefs — perceived usefulness, ease of use and compatibility — on online purchase intention; and 3) to revise the applicability of the TAM to explain Internet purchasing behavior.

In order to fulfill the objectives expounded, in the first place we develop a revision of the literature related to the Technology Acceptance Model, placing special attention on those works related to e-commerce. Moreover, we review the main contributions that support the relevance of previous experience and perceived compatibility in the adoption of Internet purchasing. On the basis of the theoretical review, research hypotheses are proposed, giving rise to an e-commerce adoption model. Subsequently, we describe the methodology developed in the research and present the empirical evidence obtained from two samples: one compounded by Internet users that have never bought anything online and the other formed by subjects with some experience with electronic transactions. Finally, we detail the most significant conclusions of the research.

2   Theoretical Background and Research Hypotheses

Below we describe the Technology Acceptance Model (TAM) and examine the relevant literature on the influence of perceived compatibility in the adoption of an innovation in general and electronic purchasing in particular.

2.1   The Technology Acceptance Model

Davis' [1] Technology Acceptance Model (TAM) is an evolution of the Theory of Reasoned Action [38], that specifically focuses on the use of new technologies. Thus, the TAM considers two specific beliefs that affect the acceptance of information systems (IS): perceived usefulness (PU) and perceived ease of use (PEOU). The first is conceptualized as the user's subjective probability that using a specific system will increase his or her performance in a particular activity, and perceived ease of use is defined as the degree to which the user expects the target system to be effortless.

The Technology Acceptance Model propounds that behavioral intention is the main determinant of the usage of an information system (Actual System Use). Additionally, the TAM includes two direct antecedents of intention to use: the individual's attitude towards the technology and its perceived usefulness [1,2]. Likewise, attitude is also influenced by perceived usefulness, while the ease of use perceived in the technology affects attitude and perceived usefulness. Figure 1 resumes the basic structure of the TAM.

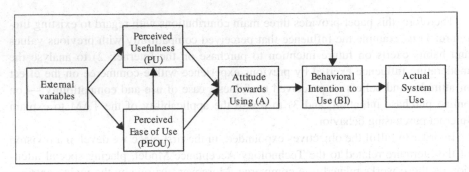

Fig. 1. Technology Acceptance Model - TAM (Source: Davis et al., 1989)

The Technology Acceptance Model is the theoretical framework which has been more extensively applied in research on information systems (IS), particularly in the scope of Internet (see [37]). In this sense, it is worth highlighting the studies which use the TAM to examine e-commerce acceptance, either considering online purchase in general [36,39-47] or examining the use of a specific service or virtual store [32,48-51]. Overall, the different works that have studied Internet behavior on the basis of the Technology Acceptance Model support the causal relationships it postulates.

Nevertheless, despite the wide support that the TAM has received in literature, there is no consensus about what its exact formulation should be. Thus, many authors have proposed alternative specifications of the model [2,52-55] adding in some cases variables not included in the original theory [40,45,56-59]. Following this approach, in this paper we intend to analyse the influence of perceived compatibility on consumers' adoption of e-commerce, incorporating this variable to the traditional formulation of the TAM.

2.2   Perceived Compatibility and Electronic Commerce Adoption

According to Rogers [7,60], perceived compatibility has been traditionally defined as "the degree to which an innovation is perceived as being consistent with the existing values, needs, and past experiences of potential adopters". More recently some authors have proposed a revised conceptualization of perceived compatibility that excludes the problematic term of "experiences", and defines this variable as the consistency of the technology with "existing values and beliefs, previously introduced ideas, and potential adopters' needs" [49]. Thus, this new definition retains the main concept of compatibility and avoids the problems derived from the lack of specificity of the meaning of the term "experiences". Anyway, different authors have supported the direct and significant relationship between user's previous experiences and the perceived compatibility with technology or behavior [61].

Within the literature on diffusion of innovations, compatibility has been traditionally considered a fundamental attribute of a new behavior or technology [6-8]. In particular, the empirical evidence obtained by Tornatzky and Klein [6] supports that this characteristic has a significant effect on the adoption of new products or systems, together with perceived usefulness and ease of use. Later on, Moore and Benbasat [9],

Agarwal and Prasad [10] and Van Slyke et al. [11] have also highlighted the importance that perceived compatibility has on the adoption of innovations. Following these studies, Taylor and Todd [56] and Agarwal and Karahanna [61] propose its incorporation in the Technology Acceptance Model.

Despite the importance attributed to perceived compatibility in the adoption of information systems, the literature about the Internet and e-commerce has paid little attention to this variable [13]. Therefore, the empirical evidence available on this topic is very limited and, additionally, contradictory. Moreover, the few studies that examine the influence of perceived compatibility on e-purchasing behavior have adopted diverse theoretical focuses and have considered different dependant variables.

In this sense, taking as a basis the Diffusion of Innovation Theory Tan and Teo [62] and Van Slyke et al. [11] support the significant influence of perceived compatibility on future intention to purchase on the Internet. With a different perspective, Agarwal and Karahanna [61], Chen et al. [49] and Herrero and Rodríguez del Bosque [12] incorporate perceived compatibility in an extended version of TAM, following the suggestions made by Taylor and Todd [56]. Thus, these authors explore the relationships between beliefs, attitudes and intentions that determine online purchasing adoption. In this sense, the empirical evidence obtained in these studies supports the influence of perceived compatibility on attitude and perceived usefulness in web use [61], in a specific website [49] and in e-commerce in general [12]. Additionally, Agarwal and Karahanna [61] observe a direct influence of perceived compatibility on perceived ease of use in the case of the web use. According to these results, and in contrast with the evidence obtained by Tan and Teo [62] and Van Slyke et al. [11], the influence of perceived compatibility on Internet purchasing behavior would be mediated through beliefs and attitudes towards the system.

In this sense, different authors [56,61] have suggested the need to examine in-depth the relationships existing among beliefs and attitudes about a system/technology and their influence on its acceptance. In accordance with this approach, in this study we consider that the influence of consumers' perceived compatibility on e-commerce acceptance is mediated through other beliefs and attitudes towards the system. Thus, taking as a basis the empirical evidence obtained by Chen et al. [49] and Herrero and Rodríguez del Bosque [12] the following hypotheses are propounded:

H1: Perceived compatibility of Internet purchasing will have a positive influence on attitude towards Internet purchasing.

H2: Perceived compatibility of Internet purchasing will have a positive influence on perceived usefulness of Internet purchasing.

H3: Perceived compatibility of Internet purchasing will have a positive influence on perceived ease of use of Internet purchasing.

2.3   Experience with e-Commerce and Future Intention to Purchase through the Internet

The experience with a conduct or product has been identified as a relevant determinant of future behavior —repetition of the conduct or continued use of the product— both in the classical theory of consumer behavior [28,29] as in the literature about attitudes formation [30,31]. In particular, there may be relevant differences between the first decision to purchase a product or service and later repetitions of this behavior. Thus, both decisions are distinct in a fundamental factor: the source of the information that generates subjects' beliefs and attitudes, which, in the end, give place to a behavior or purchase decision [30,31]. Thus, the first decision to develop a behavior or purchase a product relies on indirect information regarding the conduct or product. However, the repetition of the behavior will be affected by the subject's prior experiences with that conduct.

In accordance with this perspective, diverse authors have observed that the variables that influence the consumers' decision to make a purchase through the Internet, or the intensity of the effect exerted by these variables, may be different depending on their experience with e-commerce [13,34]. In particular, Gefen [32] observes that previous usage of a website increases future intention of use, and the perceived usefulness and ease of use perceived on the website. Gefen et al. [33] find that perceived usefulness directly influences future intention to purchase from an online store in the case of experienced users, but they do not find the same relationship for users without experience of online stores. With a similar approach, Castañeda et al. [35] find that prior experience with a website exerts a moderating effect on the determinants of the future intention to use the website. In particular, these authors observe that perceived usefulness is particularly important in the configuration of attitude towards a website in the case of individuals with higher experience of this website while ease of use is more relevant for less experienced individuals. More recently, Herrero et al. [36] have supported the moderating influence of experience with e-commerce in the relationships among beliefs and attitudes related to Internet purchasing and intention to purchase online in the future. Thus, these authors find that the influence of attitudes on online purchase intention is more intense in the case of users with prior experience with virtual transactions. Moreover, according to Herrero et al. [36] perceived usefulness directly influences the intention to purchase online for those individuals that have never purchased products on the Web, but this effect is not found for those users with prior experience of e-commerce. On the contrary, in the case of experienced users, the perceived ease of use has a relevant effect on attitudes towards e-commerce, while this influence is not supported for individuals without experience in virtual transactions. Finally, Herrero et al. [36] find that perceived ease of use exerts a significant influence on perceived usefulness for both users with and without experience of e-commerce, but this effect is stronger in the case of inexperienced individuals.

As can be observed, the empirical evidence available regarding the influence of individuals' experience of e-commerce on their online purchasing behavior is very little and contradictory —in particular, regarding the different influence of perceived usefulness and ease of use on attitudes towards online purchase between the users with and without experience of e-commerce—. Nevertheless, most studies coincide in supporting a moderating influence of experience on the relationship existing between users' beliefs and attitudes towards e-commerce and their intention to purchase online in the future. In this sense, according to the theory of attitudes formation [30,31] the relationship between beliefs, attitudes and intentions should be stronger for experienced customers because potential customers base their perceptions on indirect and relatively superficial information (see [33]). According to this perspective, and taking as a basis the widely accepted TAM, we propose the following research hypotheses:

H4: The influence of attitudes towards e-commerce on intention to purchase online will be stronger for those users with prior experience in electronic transactions than for inexperienced ones.

H5: The influence of perceived usefulness of Internet purchasing on intention to purchase online will be stronger for those users with prior experience in electronic transactions than for inexperienced ones.

H6: The influence of perceived usefulness of Internet purchasing on attitudes towards e-commerce will be stronger for those users with prior experience in electronic transactions than for inexperienced ones.

H7: The influence of perceived ease of use of Internet purchasing on attitudes towards e-commerce will be stronger for those users with prior experience in electronic transactions than for inexperienced ones.

H8: The influence of perceived compatibility of Internet purchasing on attitudes towards e-commerce will be stronger for those users with prior experience in electronic transactions than for inexperienced ones.

The proposed research hypotheses give rise to the research model which is shown in Figure 2.

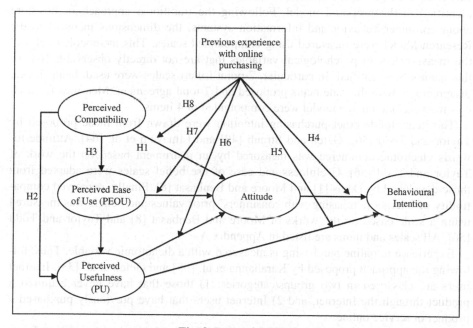

Fig. 2. Research Model

3   Research Methodology

In order to achieve the research objectives and test the hypotheses previously stated, we carried out an empirical study focused on Internet users. Firstly we developed a preliminary qualitative research which consisted of ten in in-depth interviews with

Internet experts belonging to both the professional and the academic field. The object of these interviews was to analyse the factors that determine the intention to purchase on the Internet. Given the aim of the study, we questioned the interviewees specifically about the effect exerted on the adoption of e-commerce by the perceived compatibility and previous experience with electronic transactions. The results of the qualitative research were particularly valuable in the subsequent quantitative research. Thus, they helped us to define the research population, to design the questionnaire, and to analyse and interpret the final results.

On the basis of the preliminary qualitative research, the quantitative study allowed us to determine and measure the attitudes and beliefs of Internet users with regard to e-commerce. In particular, the method chosen for the collection of the data was the personal survey. Next we describe the structure of the questionnaire used and explain in detail the research design and field work.

3.1   Questionnaire and Measures

During the field research, data was gathered through a structured questionnaire that was comprised of a set of multi-item scales which referred to the different constructs identified in the proposed model. Following the traditional approach in literature about consumer behavior and information systems, the dimensions included in the Research Model were measured using compounded scales. This methodology allows the measurement of psychological variables that are not directly observable [63] or that cannot be quantified. In particular, 7-point Likert scales were used, being 1 total disagreement with the statements proposed and 7 total agreement. Measures for each construct included in the model were compounded of 4 items.

The items for Internet purchasing intention were drawn from those proposed by Taylor and Todd [56], Gefen and Straub [48] and Limayem et al. [64]. Attitude towards electronic commerce was measured by an instrument based on the work of Taylor and Todd [56]. Usefulness and ease of use belief scales were adapted from those developed by Davis [1] and Moore and Benbasat [8]. Finally, perceived compatibility of Internet purchasing with consumers' prior values and beliefs was measured using a scale based on the works of Moore and Benbasat [8] and Taylor and Todd [56]. All scales and items are listed in Appendix A.

Experience in online purchasing is measured with a dichotomic variable. Thus, following the approach proposed by Karahanna et al. [65] and Gefen et al. [33], Internet users are classified in two groups/categories: 1) those that have never acquired a product through the Internet, and 2) Internet users that have previously purchased a product or service online.

The final questionnaire was checked by 14 academics devoted to research on Internet consumer behavior. Thus, taking as a basis a first draft of the questionnaire, the corresponding corrections were introduced attending to the suggestions made by the experts. Finally, the modified questionnaire was pre-tested on a sample of undergraduate students from a business school in order to guarantee the proper understanding of the statements proposed.

3.2   Sample and Procedure

The collection of the quantitative data was carried out through a personal survey directed to Internet users in Spain. Since it was not available a census of Internet users in the area of research, we used a convenience sampling procedure (non-probabilistic). Thus, individuals were sampled in main commercial malls and streets, and were polled if they used the Internet at least half an hour in a regular week (the minimum use rate that defines an Internet user according with the results of the previous qualitative study). Additionally, with the aim of obtaining a sample as representative as possible we imposed a sample stratification according to the profile of Internet users in Spain, defined by the Estudio General de Medios (EGM). Specifically, the sample was stratified in accordance with two users' characteristics: age and gender.

1,008 surveys to Internet users were obtained in the research work, of which 10 were eliminated since they were incomplete or had anomalous responses. 675 (67.64 %) out of the 998 complete surveys correspond to users that have never bought on the Internet and 323 (32.36 %) to individuals that have purchased products from the Internet prior to the performance of the survey. Table 1 resumes the socio-demographic characteristics of Internet users in Spain (EGM), and the profile of the two samples considered in our research.

Table 1. Socio-demographic profile of the population and samples

Variable	Population - EGM (%)	Overall sample (%)	Sample 1 – Non-buyers (%)	Sample 2 – Buyers (%)
Gender				
Male	57.4	56.9	53.0	65.0
Female	42.6	43.1	47.0	35.0
Age				
14 to 19 years	17.2	16.9	20.0	10.5
20 to 24 years	18.3	19.8	18.8	22.0
25 to 34 years	31.0	31.1	29.1	35.0
35 to 44 years	19.5	18.4	16.3	22.9
45 to 54 years	10.0	9.5	10.5	7.4
55 and over	4.0	4.3	5.3	2.2
Education level				
Less than primary	0.5	0.6	0.8	0.3
Primary	25.2	12.8	16.7	4.5
Secondary	39.1	39.0	38.1	40.8
College	35.2	47.6	44.4	54.4
Marital status				
Single	57.7	63.1	63.2	62.7
Married	39.9	34.9	34.5	36.0
Divorced	1.9	1.5	1.6	1.3
Widow/er	0.5	0.5	0.7	0.0
Occupation				
Employed	61.0	52.0	48.3	59.8
Unemployed	39.0	48.0	51.7	41.2

As can be observed in Table 1 the sample obtained in this research is very similar to the profile of Internet users in Spain (EGM). Besides, some relevant differences between the two samples considered are shown. In particular, in the case of users with previous experience in electronic transactions there is a higher proportion of males (65% against 53%), of university graduates (54.4% against 44.4%) and employed people (59.8% against 48.3%).

4   Results

With the aim of testing the research model propounded, the hypotheses are firstly tested independently for each one of the two samples considered (Internet users with and without prior experience with online transactions). In particular, we follow a Structural Equations Model (SEM) approach with Maximum Likelihood Robust estimation, using the software package EQS 6.1. Thus, a Confirmatory Factor Analysis (CFA) is first performed in order to test the psychometric properties of the measurement scales for all the variables considered in the research model. Then, a causal model of structural equations is estimated for each sample. Finally, for the two samples considered the estimated coefficients are compared to analyse the moderating influence exerted by previous experience of e-commerce.

4.1   Validity and Reliability of Measurement Instruments

Regarding the psychometric properties of the scales, for the two samples studied, the initial Confirmatory Factor Analyses show the need to eliminate item PEOU3 from the scale of perceived ease of use, since the corresponding standardized lambda coefficients are lower than the required minimum value of 0.5. Once this correction is done, the values obtained for Goodness of fit indices[1] confirm the appropriate specification of the factorial structure propounded (Table 2). Thus, the Bentler-Bonett Normed Fit Index (BBNFI), Bentler-Bonett Not Normed Fit Index (BBNNFI) and Incremental Fit Index (IFI) take values above the recommended minimum of 0.9. Similarly, the Root Mean Square Error of Approximation (RMSEA) is within the maximum recquired limit of 0.08 [66]. Besides, the Normed $\chi2$ statistic is below the recommended value of 5.0, thus confirming parsimony fit [66].

Additionally, the convergent validity of the measurement instruments is confirmed for the two samples (Table 2). Thereby, all items are significant to a confidence level of 95% and the standardized lambda coefficients are higher than 0.5 [67]. Discriminant validity of the scales is tested according with the procedure described by Anderson and Gerbing [68]. With this aim, we examine if the confidence intervals for the

---

[1] The fit criteria of a structural equation model indicate to what extent the specified model fits the empirical data. Three main classes of criteria exist: measures of absolute fit, measures of incremental fit, and measures of parsimonious fit [66]. In this case we use the statistics given by EQS 6.1, widely used in SEM literature [66][72]: BBNFI, BBNNFI and RMSEA for the measurement of overall model fit, IFI as measure of incremental fit and Normed $\chi2$ for the measurement of parsimony of the model.

correlation between constructs contains the value 1. The estimated intervals do not contain the unit in any of the cases, thus supporting the discriminant validity of the measurement model proposed (Appendix B).

Cronbach's Alpha (Cronbach, 1951), compound reliability and AVE coefficients [69] are used to evaluate the inner reliability of the measurement instruments. In every case, these statistics are above the required minimum values of 0.7 and 0.5 respectively [66,69], thus supporting the reliability of the constructs for the two samples studied (Table 2).

Table 2. Confirmatory Factor Analyses

Latent variable	Latent variable	Measured variable	Standar. lambda	R^2	Cronbach's α	Composite reliability	Goodness of fit indices
Sample 1 – Internet users without experience in online purchasing	Intention	INTEN1	0.902	0.814	0.879	0.901	Normed χ^2 = 3.31 BBNFI = 0.946 BBNNFI = 0.948 IFI = 0.957 RMSEA = 0.073
		INTEN2	0.940	0.884			
		INTEN3	0.879	0.773			
		INTEN4	0.575	0.331			
	Attitude	ATTIT1	0.806	0.650	0.943	0.945	
		ATTIT2	0.906	0.820			
		ATTIT3	0.954	0.910			
		ATTIT4	0.932	0.869			
	Perceived usefulness	PU1	0.818	0.669	0.895	0.895	
		PU2	0.787	0.619			
		PU3	0.875	0.765			
		PU4	0.816	0.666			
	Perceived ease of use	PEOU1	0.902	0.813	0.903	0.906	
		PEOU2	0.942	0.888			
		PEOU4	0.769	0.591			
	Perceived compatibility	COMP1	0.907	0.822	0.940	0.941	
		COMP2	0.930	0.865			
		COMP3	0.885	0.783			
		COMP4	0.853	0.728			
Sample 2 – Internet users with previous experience in online purchasing	Intention	INTEN1	0.900	0.811	0.861	0.866	Normed χ^2 = 1.95 BBNFI = 0.920 BBNNFI = 0.936 IFI = 0.947 RMSEA = 0.075
		INTEN2	0.940	0.884			
		INTEN3	0.783	0.613			
		INTEN4	0.470	0.221			
	Attitude	ATTIT1	0.851	0.724	0.937	0.939	
		ATTIT2	0.883	0.779			
		ATTIT3	0.945	0.893			
		ATTIT4	0.887	0.787			
	Perceived usefulness	PU1	0.749	0.560	0.839	0.840	
		PU2	0.675	0.455			
		PU3	0.829	0.687			
		PU4	0.756	0.571			
	Perceived ease of use	PEOU1	0.830	0.689	0.793	0.810	
		PEOU2	0.855	0.732			
		PEOU4	0.597	0.356			
	Perceived compatibility	COMP1	0.875	0.765	0.947	0.949	
		COMP2	0.917	0.841			
		COMP3	0.945	0.892			
		COMP4	0.888	0.789			

4.2   Perceived Compatibility in Online Purchasing (Structural Model Estimation)

Once the psychometric properties of the measurement instruments are contrasted, the structural model propounded to explain e-commerce acceptance is estimated for each sample studied (Internet users with and without previous experience in online transactions). A first estimation of the causal model by the Maximum Likelihood Robust method shows the existence of significant differences between both samples considered. These differences involve both the causal relationships included in the Technology Acceptance Model [1,2] and the influence of perceived compatibility on Internet purchasing behavior.

With respect to the relationships included in the Technology Acceptance Model [1,2], for those individuals that have already made purchases on the Internet, all causal effects propounded in the TAM are supported. However, for the sample of Internet users without previous experience of online transactions a significant influence of perceived usefulness of the system on the intention to purchase online is not found, neither is the perceived ease of use on attitudes towards e-commerce.

Additionally, for those Internet users without previous experience of online purchasing, the LM Test suggests the existence of a causal effect not considered in the initial model: a direct and positive effect of perceived compatibility on Internet purchase intention. Although this causal relationship is not included in the original research model, it is consistent with the Diffusion of Innovation Theory and is supported by the empirical evidence obtained by Tan and Teo [62] and Van Slyke et al. [11]. Thus, following the approach proposed by Hair et al. [66] incorporating this causal link in the structural model is acceptable (and recommendable), as it is appropriately supported by both empirical evidence and theoretical basis.

In accordance with these results, the causal model proposed is completely supported for the sample of Internet users that have already purchased online, but some modifications seem needed for the sample of Internet users without previous experience of virtual purchases. Therefore, following the model development strategy [66], the structural model is respecified for that sample, removing the non-significant relationships (influence of perceived usefulness on intention and of perceived ease of use on attitudes towards e-commerce), and introducing a direct and positive effect of perceived compatibility on Internet purchase intention. Accordingly, the goodness of fit indices indicate an appropriate specification of the original research model for those individuals that have previously purchased through the Internet, and of the respecified model for the sample of Internet users without previous experience of online purchases (Table 3).

Table 3. Structural Models - Goodness of Fit Indices

	BBNFI	BBNNFI	IFI	RMSEA	Normed χ^2
Sample 1 – Internet users without experience in online purchasing	0.945	0.949	0.957	0.072	3.36
Sample 2 – Internet users with previous experience in online purchasing	0.918	0.934	0.945	0.076	2.01

Table 4 shows the results of the structural models estimation for both samples considered: users with and without previous experience of online purchases.

Regarding the influence of perceived compatibility on Internet purchasing behavior, the empirical results obtained support H1, H2 and H3 (Table 4). Thus, significant effects of this variable on attitudes towards e-commerce and perceived usefulness and ease of use in the system are found for both Internet users with and without previous experience of virtual transactions. Therefore, to the extent that individuals find e-commerce to be compatible with their values and previous habits, their beliefs about the usefulness and ease of use of Internet purchasing and their attitudes towards the system will become stronger. Additionally, in the case of those users without previous experience of virtual transactions, perceived compatibility of e-commerce exerts a direct influence on Internet purchasing intention.

Table 4. Structural Models Estimation

	Sample 1 – Internet users without experience in online purchasing		Sample 2 – Internet users with previous experience in online purchasing	
	R^2	Standard. Coeff.	R^2	Standard. Coeff.
Purchase Intention	0.483		0.449	
Attitude		0.357**		0.493**
Perceived Usefulness		n.s.		0.220**
Perceived Compatibility		0.418**		-
Attitude	0.552		0.582	
Perceived Usefulness		0.554**		0.491**
Perceived Ease of use		n.s.		0.144**
Perceived Compatibility (H1)		**0.257****		**0.255****
Perceived Usefulness	0.470		0.469	
Perceived Ease of use		0.293**		0.225**
Perceived Compatibility (H2)		**0.548****		**0.564****
Perceived Ease of use	0.068		0.156	
Perceived Compatibility (H3)		**0.261****		**0.395****

** p-value < 0.05.

4.3   Previous Experience of Online Purchasing (Multi-sample Analysis)

With respect to the moderating influence of previous experience of virtual transactions on online purchasing behavior (H4 to H8), some conclusions can be directly extracted from the estimation of the structural models for both samples considered. Thus, while all causal relationships considered in the TAM are confirmed for those individuals with previous experience of online purchasing, some of them are not significant for Internet users that have not made virtual purchases in the past (Table 4). In particular, as has been stated above, for this last group, the effects of perceived usefulness on intention to purchase online and of the perceived ease of use on attitudes towards e-commerce are not statistically significant. Therefore, hypotheses H5 and H7 are supported, since there are evident differences between each sample regarding these relationships.

With regard to hypotheses H4, H6 and H8, no conclusions can be directly obtained from the structural models estimation, given that the causal relationships propounded are confirmed for both samples considered. In this case, it is necessary to examine the intensity of each of these relationships for Internet users with and without previous experience of online purchasing. In particular, the comparison between the estimated

Table 5. Comparison of the samples of Internet users without and with experience in online transactions (unstandardized coefficients)

Causal relationship	Sample 1 – Internet users without experience in online purchasing	Sample 2 – Internet users with previous experience in online purchasing	T Statistical
H4: Attitude → Intention	**0.393**	**0.593**	**-2.005****
H5: Perceived Usefulness → Intention	**n.s.**	**0.321**	**-**
H6: Perceived Usefulness → Attitude	0.549	0.595	-0.454
H7: Perceived Ease of Use → Attitude	**n.s.**	**0.165**	**-**
H8: Perceived Compatibility → Attitude	0.257	0.217	0.540

** Differences between unstandardized coefficients for each sample are significant at p-value <0.05.

coefficients for both groups and each pair of variables (Table 5) was carried out using a measurement of the signification of the differences between coefficients using a t-test for independent samples (see [35,70]).

In accordance with the results summarized in Table 5 hypothesis H4 is supported, while hypotheses H6 and H8 are not supported. Thus, the influence of perceived usefulness and perceived compatibility on attitudes towards e-commerce is statistically significant and of similar intensity for both the Internet users without previous experience of online transactions and for those individuals that have already made purchases on the Internet. On the contrary, the effect exerted by attitudes towards e-commerce on intention to purchase on the Internet in the future is significantly stronger for users with previous experience of online transactions (H4).

Finally, regarding the moderating influence of previous experiences of e-commerce on online purchasing behavior, another relevant and unexpected result must be highlighted. In particular, according to the empirical evidence detailed above (Table 4), the perceived compatibility with e-commerce exerts a direct and positive influence on online purchasing intention for those users without previous experience of transactions on the Internet. Although this causal relationship is not included in the initial research model, it is supported both by the statistical results and by the literature. Therefore, a direct influence of perceived compatibility on acceptance intention is consistent with the Diffusion of Innovation Theory and is supported by the empirical evidence obtained by Tan and Teo [62] and Van Slyke et al. [11].

In accordance with this result, the influence of perceived compatibility of e-commerce with individual's previous habits and values seems to be particularly relevant in the decision to make a first purchase on the Internet (initial acceptance), and could explain other unexpected results obtained (the non-significant effect of perceived usefulness on attitudes towards e-commerce and of perceived ease-use on perceived usefulness). This issue is discussed in depth in the following section.

5   Discussion

Next we explain the most important conclusions of our research as well as its main implications for the management of e-commerce firms and the related organizations. To conclude, we highlight the limitations of our study and future research lines.

5.1 Conclusions

This paper analyses the factors that determine the adoption of electronic commerce by consumers and that lead to Internet users becoming online purchasers. In particular, we examine in depth the influence of two variables that have been identified as determinants of online purchasing behavior, but that have rarely been studied in this field: perceived compatibility of e-commerce with previous values and habits, and prior experience of online purchasing. With this aim, the Technology Acceptance Model –TAM– [1,2] is taken as a reference framework. The selection of this theoretical basis is justified by its extensive use in the literature about information systems acceptance and, particularly, in the field of Internet (see [37]). Therefore, we propose an extended model of online purchase, adding the perceived compatibility as a causal determinant of the variables included in the TAM (beliefs and attitudes about technology), and previous experience of e-commerce as a moderator variable in the relationship between beliefs, attitudes and intention. Thus, in order to examine the influence exerted by previous experience on online purchasing behavior, the research sample has been divided between those Internet users that have never purchased on the Internet and those that have already purchased online in the past.

Regarding the influence of perceived compatibility of e-commerce with users' previous values and habits, the empirical evidence obtained supports that this variable is a significant determinant of online purchasing for the two groups considered in the study. Thus, perceived compatibility positively influences attitudes towards e-commerce and perceived usefulness and ease of use of online purchasing for Internet users with and without previous experience of virtual transactions. Therefore, more compatibility with their previous habits and values perceived by individuals in e-commerce will give place to more positive beliefs and attitudes with regard to online purchasing.

With respect to the moderating influence of previous experience of virtual transactions on online purchasing behavior, the results obtained in this research indicate the existence of significant differences between users with and without previous experience of online purchasing. These differences involve both the causal relationships included in the Technology Acceptance Model [1,2] and the influence of perceived compatibility on Internet purchasing behavior.

Regarding the TAM, the empirical evidence obtained shows that this theoretical framework would not be equally applicable for explaining the first adoption of Internet purchasing and the subsequent repetition of virtual purchases. Thus, although the TAM is entirely valid to explain online purchasing intention for those users with experience of virtual transactions (all the relationships included in the model are statistically significant), it is not completely supported for inexperienced users. Therefore, in the case of those users that have not made virtual transactions in the past, perceived usefulness only influences Internet purchasing intention indirectly, through its effect on attitudes towards e-commerce. Similarly, for this group of users, perceived ease of use in online purchases only influences perceived usefulness in the system, but not the attitudes towards e-commerce. Moreover, it is observed that the influence of attitudes towards e-commerce on intention to purchase on the Internet is more intense for those users that have previously purchased online. These results can be explained by the theory of attitude formation [30,31], which states that direct experience of a behavior (e.g. online purchasing) reinforces individuals' beliefs about the conduct and gives rise to more consistent and structured relationships between beliefs, attitudes and behavioral intention.

Finally, according to the results obtained, perceived compatibility of e-commerce with previous habits and values exerts a direct influence on intention to purchase on-line in the future for those users without previous experience in virtual transactions. Therefore, the influence of perceived compatibility seems to be more significant in the initial adoption of Internet purchasing than in later repetitions of this behavior. This phenomenon can be explained by the characteristics of the medium itself, which has a heavy technological component. Thus, given the variety of alternative shopping channels, consumers will begin to purchase on the Internet only if they feel comfortable and safe with the medium and perceive electronic commerce as coherent with their values and lifestyle. Otherwise, individuals will continue using traditional channels, more adapted to and integrated with their former experiences and habits.

On the contrary, in the case of those consumers that have already bought some products online, the effect of perceived compatibility on purchase intention is found to be weaker and exerted indirectly through attitudes and perceived usefulness associated with e-commerce. Thus, once consumers have made some transactions online, perceived compatibility will be high, as a result of the direct contact with and knowledge about the system. However, direct experience with e-commerce does not lead to such an intense variation in intentions and attitudes towards the behavior. Correlations between perceived compatibility and Internet buying intention will therefore weaken, and behavioral intention will be explained mainly by attitude towards the system and perceived usefulness. In this sense, individuals will only continue buying on the Internet if their prior purchases have been satisfactory enough and have reinforced their attitudes towards e-commerce and the usefulness perceived in this behavior. Otherwise, and despite considering Internet buying totally compatible with their prior values and habits, consumers will not carry out new transactions online.

5.2   Management Implications

From a practical perspective, the results obtained can be highly useful for those firms and professionals working in the field of e-commerce. In particular, a deep understanding of the process that determines consumers acceptance of e-commerce is fundamental to take efficient marketing decisions aimed to encourage electronic transactions and foster the growth of virtual markets. Given the fundamental role of attitudes on online purchase acceptance, the first efforts of firms and public administrations should focus on improving the opinion and perceptions of the whole society with regard to e-commerce. Thus, e-commerce companies should not direct their marketing strategies only to improve their brand image and increase their client portfolio, but they must also try to foster e-commerce acceptance in general. Therefore, contributing to the overall diffusion of Internet purchases is the only way to promote future growth of the business. That is to say, the firms must work in building a bigger electronic market and not only in gaining market share.

Besides, individuals' perceptions about the attributes of online purchasing, and particularly its usefulness, ease of use and compatibility, have a relevant influence in their acceptance of e-commerce and their attitudes toward this behavior. Accordingly, e-commerce companies must communicate and highlight the advantages and superior value provided by their web pages and electronic stores compared with traditional channels. Moreover, it is very important to offer consumers simple and friendly shopping processes, providing all the possible facilities to search for products, confirm the order, make the payment and follow the delivery.

Finally, the identification of two clearly differentiated stages in the adoption process (initial decision to make a first purchase and the later repetition of the behavior), in which Internet purchasing intention is influenced by different variables and with different intensity, raises the need to develop different strategies to attract first users and consolidate of experienced ones. Therefore, in order for Internet users to become buyers, web-based sellers should make an effort to develop consumers' knowledge and familiarity with virtual environments and online transaction process (that is to say, increase perceived compatibility in e-commerce of those users that have never purchased on the Internet in the past). On the other hand, once individuals have acquired enough experience with Internet purchases, sellers should highlight those attributes of e-commerce linked more to utilitarian and practical issues. Thus, unless consumers have a positive attitude towards online purchasing and perceive this behavior as useful, they will use other alternative channels (even despite the fact that they perceive e-commerce as compatible with their values and habits).

5.3   Limitations and Future Research Lines

Despite the systematic and thorough methodology followed in this research, it presents some limitations that must be pointed out. Firstly, this study focuses in e-commerce acceptance in general, and thus examines users' attitudes, beliefs and intentions regarding online purchasing of any product, without taking into consideration a particular goods or services category. This fact could affect the results obtained in the research, since different authors claim that not every product has the same potential for online selling. This research focuses on e-commerce in general because our objective is to propound an overall model to explain online purchase, unaffected by the particular characteristics of a specific product category. Nevertheless, the analysis of the effect exerted by the typology of product purchased and its attributes is a very interesting line for future research. In this sense, it would be of particular relevance the study of the differences existing between the Internet purchase of goods and services.

Likewise, regarding the methodology of research, the endogenous variable considered in this study might also be a limitation. Thus, the consumer intention to purchase online is measured subjectively, by means of the individuals' perceptions regarding they future behavior [53,56]. Although this method has been adopted in many studies, some researchers suggest that some other behavior-oriented measure such as choice behavior should be used instead [54]. Thompson, Higgins and Howell [71] further postulate that both objective and subjective measures should be employed and that the correspondence between them should be examined. Accordingly, another line for future research would be to focus on the analysis of the coincidence between online purchase intention and the actual behavior.

The results obtained in this paper have some interesting implications for future research. Thus, the studies developed confirm the value of the TAM as a theoretical framework to explain both the decision to purchase on the Internet for the first time and the repetition of this behavior by those consumers with prior experience with online transactions. However, this theoretical framework seems not to be equally applicable in both situations. Therefore, more empirical evidence seems needed to clarify how attitudes, and perceived usefulness and ease of use influence purchase online decision depending of the individuals' previous experience with this behavior. Moreover, it would be very interesting to deepen in the analysis of the different stages of e-commerce adoption process, and the study of the relevant variables in each case. Thus, while we have distinguished between Internet users without and with previous experience in online purchasing, it could be possible to find more accurate stages based on a more specific analysis of consumers' experience, such as the amount of time spent buying online or number of transactions made, among others. Additionally, it would be necessary to obtain new empirical evidence that confirms the influence of perceived compatibility on Internet purchasing behavior. Likewise, future research should analyse the variables that could condition the effect of perceived compatibility on intention and attitudes towards online purchasing. For example, these links could be moderated by the type of product bought –services versus tangible goods–, its relative value –high cost versus low cost products–, or consumer personality traits, like personal innovativeness or risk taking propensity.

References

1. Davis, F.D.: Perceived usefulness, perceived ease of use, and user acceptance of information technology. MIS Quart. 13, 319–339 (1989)
2. Davis, F.D., Bagozzi, R.P., Warshaw, P.R.: User Acceptance of Computer-Technology - a Comparison of 2 Theoretical-Models. Manage. Sci. 35, 982–1003 (1989)
3. Jones, J.M., Vijayasarathy, L.R.: Internet consumer catalog shopping: Findings from an exploratory study and directions for future research. Internet. Res. 8, 322–330 (1998)
4. Goldsmith, R.E., Bridges, E.: E-Tailing vs. Retailing. Using Attitudes to Predict Online Buying Behavior. Quart. J. Electron. Comm. 1, 245–253 (2000)
5. Rowley, J., Slack, F.: Leveraging customer knowledge – Profiling and personalisation in e-business. Int. J. Ret. & Distrib. Manag. 29, 409–416 (2001)
6. Tornatzky, L.G., Klein, K.J.: Innovation Characteristics and Innovation Adoption-Implementation: A Meta-Analysis of Findings. IEEE Trans. Eng. Manage. EM-29, 28–45 (1982)
7. Rogers, E.M.: Diffusion of Innovations, 3rd edn. Free Press, Collier Macmillan, New York (1983)
8. Moore, G.C., Benbasat, I.: Development of an instrument to measure the perceptions of adopting an information technology innovation. Inf. Syst. Res. 2, 192–222 (1991)
9. Moore, G.C., Benbasat, I.: Integrating Diffusion of Innovations and Theory of Reasoned Action Models to Predict Utilization of Information Technology by End-Users. In: Kautz, K., Pries-Heje, J. (eds.) Diffusion and Adoption of Information Technology, pp. 132–146 (1996)

10. Agarwal, R., Prasad, J.: The role of innovation characteristics and perceived voluntariness in the acceptance of information technologies. Decision Sci. 28, 557–579 (1997)

11. Van Slyke, C., Belanger, F., Comunale, C.L.: Factors influencing the adoption of web-based shopping: The impact of trust. Data. Base. Adv. Inf. Syst. 35, 32–47 (2004)

12. Crespo, A.H., Rodriguez, I.A.R.D.: Explaining B2C e-commerce acceptance: An integrative model based on the framework by Gatignon and Robertson. Interact. Comput. 20, 212–224 (2008)

13. Chang, M.K., Cheung, W., Lai, V.S.: Literature derived reference models for the adoption of online shopping. Inf. Manag. 42, 543–559 (2005)

14. Dahlen, M.: Closing in on the web consumer - A study in Internet shopping. Conv. Comm. Bey., 121–137 (2000)

15. Eastlick, M.A., Lotz, S.: Profiling potential adopters and non-adopters of an interactive electronic shopping medium. Int. J. Ret. & Distrib. Manag. 27, 209–223 (1999)

16. Miyazaki, A.D., Fernandez, A.: Consumer perceptions of privacy and security risks for online shopping. J. Consumer Aff. 35, 27–44 (2001)

17. Dholakia, R.R., Uusitalo, O.: Switching to electronic stores: consumer characteristics and the perception of shopping benefits. Int. J. Ret. & Distrib. Manag. 30, 459–469 (2002)

18. Li, H., Kuo, C., Russell, M.G.: The impact of perceived channel utilities, shopping orientations, and demographics on the consumer's online buying behavior. J. Comput.-Mediat. Commun. 5 (1999)

19. Lohse, G.L., Bellman, S., Johnson, E.J.: Consumer buying behavior on the internet: Findings from panel data. J. Interact. Mark. 14, 15–29 (2000)

20. Flynn, L.R., Goldsmith, R.E.: The Impact of Internet Knowledge on Online Buying Attitudes, Behavior and Future Intentions: A Structural Modelling Approach. Soc. Market. Adv. Proceed., 193–196 (2001)

21. Kwak, H., Fox, R.J., Zinkhan, G.M.: What products can be successfully promoted and sold via the internet? J. Advert. Res. 42, 23–38 (2002)

22. Bellman, S., Lohse, G.L., Johnson, E.J.: Predictors of online buying behavior. Commun. ACM 42, 32–38 (1999)

23. Van Den Poel, D., Leunis, J.: Consumer acceptance of the internet as a channel of distribution. J. Bus. Res. 45, 249–256 (1999)

24. Bhatnagar, A., Misra, S., Rao, H.R.: On risk, convenience, and internet shopping behavior. Commun. ACM 43, 98–105 (2000)

25. Citrin, A.V., Sprott, D.E., Silverman, S.N., Stem Jr., D.E.: Adoption of Internet shopping: The role of consumer innovativeness. Ind. Manag. Dat. Syst. 100, 294–300 (2000)

26. Liao, Z., Cheung, M.T.: Internet-based e-shopping and consumer attitudes: An empirical study. Inf. Manag. 38, 299–306 (2001)

27. Park, C., Jun, J.K.: A Cross-Cultural Comparison of Online Buying Intention: Effects of Internet Usage, Perceived Risk, and Innovativeness. Int. Market. Rev. 20, 534–553 (2003)

28. Howard, J.A., Sheth, J.N.: The Theory of Buyer Behavior. John Wiley & Sons, New York (1969)

29. Engel, J.F., Blackwell, R.D., Miniard, P.W.: Consumer Behaviour. Harcourt Education, Port Melbourne (1995)

30. Ajzen, I.: The theory of planned behavior. Organ. Behav. Hum. Decis. Process. 50, 179–211 (1991)

31. Ajzen, I., Fishbein, M.: The influence of attitudes on behavior. In: Albarracín, D., Johnson, B.T., Zanna, M.P. (eds.) The Handbook of Attitudes, pp. 173–221. Erlbaum, Mahwah (2005)

32. Gefen, D.: TAM or just plain habit: A look at experienced online shoppers. J. End User Comput. 15, 1–13 (2003)
33. Gefen, D., Karahanna, E., Straub, D.W.: Inexperience and experience with online stores: The importance of TAM and trust. IEEE Trans. Eng. Manage. 50, 307–321 (2003)
34. Hsu, M.-., Yen, C.-., Chiu, C.-., Chang, C.-.: A longitudinal investigation of continued online shopping behavior: An extension of the theory of planned behavior. Int. J. Hum.-Comput. Stud. 64, 889–904 (2006)
35. Castañeda, J.A., Muñoz-Leiva, F., Luque, T.: Web Acceptance Model (WAM): Moderating effects of user experience. Inf. Manag. 44, 384–396 (2007)
36. Crespo, A.H., del Bosque, I.R., Sanchez, M.M.G.D.: The influence of perceived risk on Internet shopping behavior: a multidimensional perspective. J. Risk. Res. 12, 259–277 (2009)
37. Lee, Y., Kozar, K.A., Larsen, K.R.T.: The Technology Acceptance Model: Past, Present, and Future. Commun. AIS 12, 752–780 (2003)
38. Ajzen, I., Fishbein, M.: Understanding Attitudes and Predicting Social Behavior. Prentice Hall, Englewood Cliffs (1980)
39. Teo, T.S.H., Lim, V.K.G., Lai, R.Y.C.: Intrinsic and extrinsic motivation in Internet usage. Omega 27, 25–37 (1999)
40. Childers, T.L., Carr, C.L., Peck, J., Carson, S.: Hedonic and utilitarian motivations for online retail shopping behavior. J. Retail. 77, 511–535 (2001)
41. Fenech, T., O'Cass, A.: Internet Users' Adoption of Web Retailing: User and Product Dimensions. J. Prod. Brand. Manage. 10, 361–381 (2001)
42. Salisbury, W.D., Pearson, R.A., Pearson, A.W., Miller, D.W.: Perceived security and World Wide Web purchase intention. Ind. Manag. Dat. Syst. 101, 165–176 (2001)
43. O'Cass, A., Fenech, T.: Web retailing adoption: Exploring the nature of internet users Web retailing behaviour. J. Retail. Consum. Serv. 10, 81–94 (2003)
44. Park, J., Lee, D., Ahn, J.: Risk-Focused e-Commerce Adoption Model: A Cross-Country Study. J. Glob. Inform. Tech. Manag. 7, 6–30 (2004)
45. Shih, H.-.: An empirical study on predicting user acceptance of e-shopping on the Web. Inf. Manag. 41, 351–368 (2004)
46. Bosnjak, M., Obermeier, D., Tuten, D.: Predicting and explaining the propensity to bid in online auctions - A comparison of two action-theoretical models. J. End User Comput. 5, 102–116 (2006)
47. Rodriguez, I.A.R.D., Crespo, A.H.: Antecedentes de la utilidad percibida en la adopción del comercio electrónico entre particulares y empresas. Cuad. Econ. Direc. Emp. 34, 107–134 (2008)
48. Gefen, D., Straub, D.W.: The Relative Importance of Perceived Ease of Use in IS Adoption: A Study of E-Commerce Adoption. J. Assoc. Inf. Syst. 1 (2000)
49. Chen, L.-., Gillenson, M.L., Sherrell, D.L.: Enticing online consumers: An extended technology acceptance perspective. Inf. Manag. 39, 705–719 (2002)
50. Pavlou, P.A.: Consumer acceptance of electronic commerce: Integrating trust and risk with the technology acceptance model. Int. J. Electron. Commer. 7, 101–134 (2003)
51. Heijden, H.V.D., Verhagen, T., Creemers, M.: Understanding online purchase intentions: Contributions from technology and trust perspectives. Eur. J. Inform. Syst. 12, 41–48 (2003)
52. Adams, D.A., Nelson, R.R., Todd, P.A.: Perceived usefulness, ease of use, and usage of information technology: A replication. MIS Quart. 16, 227–247 (1992)
53. Chau, P.Y.K.: An Empirical Assessment of a Modified Technology Acceptance Model. J. Manage. Inf. Syst. 13, 185–204 (1996)

54. Szajna, B.: Empirical evaluation of the revised technology acceptance model. Manage. Sci. 42, 85–92 (1996)
55. Venkatesh, V., Davis, F.D.: Theoretical extension of the Technology Acceptance Model: Four longitudinal field studies. Manage. Sci. 46, 186–204 (2000)
56. Taylor, S., Todd, P.A.: Understanding information technology usage: A test of competing models. Inf. Syst. Res. 6, 144–176 (1995)
57. Venkatesh, V.: Creation of favorable user perceptions: Exploring the role of intrinsic motivation. MIS Quart. 23, 239–260 (1999)
58. Bhattacherjee, A.: An empirical analysis of the antecedents of electronic commerce service continuance. Decis. Support Syst. 32, 201–214 (2001)
59. Featherman, M.S., Pavlou, P.A.: Predicting e-services adoption: A perceived risk facets perspective. Int. J. Hum.-Comput. Stud. 59, 451–474 (2003)
60. Rogers, E.M.: Diffusion of Innovations, 4th edn. Free Press, New York (1995)
61. Agarwal, R., Karahanna, E.: On the Multi-dimensional Nature of Compatibility Beliefs in Technology Acceptance. Proc. DIGIT Conf. (1998)
62. Tan, M., Teo, T.S.H.: Factors influencing the adoption of Internet banking. J. Assoc. Inf. Syst. 1 (2000)
63. Churchill, G.A., Iacobucci, D.: Marketing Research: Methodological Foundations, 8th edn. Harcourt College, Fort Worth, Texas (2002)
64. Limayem, M., Khalifa, M., Frini, A.: What makes consumers buy from Internet? A longitudinal study of online shopping. IEEE Trans. Syst. Man. Cybern. 30, 421–432 (2000)
65. Karahanna, E., Straub, D.W., Chervany, N.L.: Information technology adoption across time: A cross-sectional comparison of pre-adoption and post-adoption beliefs. MIS Quart. 23, 183–213 (1999)
66. Hair, J.F., Anderson, R.E., Tatham, R.L., Black, W.C.: Multivariate Data Analysis, 5th edn. Prentice-Hall (1998)
67. Steenkamp, J.-.E.M., van Trijp, H.C.M.: The use of lisrel in validating marketing constructs. Int. J. Res. Mark. 8, 283–299 (1991)
68. Anderson, J.C., Gerbing, D.W.: Structural Equation Modeling in Practice: A Review and Recommended Two-Step Approach. Psychol. Bull. 103, 411–423 (1988)
69. Nunally, J.C.: Psychometric Theory, 2nd edn. McGraw-Hill (1978)
70. Lee, J., Kim, J., Moon, J.Y.: What makes internet users visit cyber stores again? Key design factors for customer loyalty. CHI Letters 2, 305–312 (2000)
71. Thompson, R.L., Higgins, C.A., Howell, J.M.: Influence of experience on personal computer utilization: Testing a conceptual model. J. Manage. Inf. Syst. 11, 167–187 (1994)
72. Byrne, B.M.: Structural Equation Modeling with EQS and EQS/WINDOWS. Basic Concepts, Applications, and Programming. Sage Publications, Inc. (1994)

Behavior Analysis of Video Hosting Website Users Based on an Extended Technology Acceptance Model

Ayako Hiramatsu and Kazuo Nose

Osaka Sangyo University, Osaka, Japan
{ayako,nose}@ise.osaka-sandai.ac.jp

Abstract. This paper shows an analysis why students use online video hosting services. Based on the Technology Acceptance Model (TAM), a hypothesis model for video hosting website users was designed. The hypothesis model incorporates social influence, flow experience, comfortable communication, and advertisement interference into the basic TAM. A questionnaire was given to about 350 students. Behaviors of online video hosting website users were analyzed with correlation analysis and structural equation modeling.

Keywords: Video Hosting Service, Technology Acceptance Model, User Behavior.

1 Introduction

The Internet penetration rate in Japan was 78.2% in 2010 and the number of Internet users was estimated at 94.62 million [1]. With Internet penetration, services using the Internet tend to increase. One such service is online video. There are two main types of online video services: Internet television and video hosting. Users of both services are increasing: in particular, YouTube is the most famous video hosting service worldwide. Online video services are expected to offer effective advertising for users who to watch video content as well as providing video content. There is the possibility of expanding online video services as information distributors that can introduce video content and related products. For this reason, we consider it important to under-stand why people use online video hosting services.

User models are often studied to understand user behaviors by analyzing behavior data. To understand user behavior in online shopping, there are studies that analyze factors affecting consumer decisions for purchasing PCs [2], and modeling the decision-making process for online music services [3,4]. In regard to information systems, the Technology Acceptance Model (TAM) [5] is often applied for user modeling research. There are many studies based on TAM for services using the Internet, such as online shopping [6,7,8], mobile TV [9], Skype-Out [10], and mobile learning[11]. These are newly introduced factors to basic TAM and proposed extended TAMs.

T. Matsuo & R. Colomo-Palacios (Eds.): *Electronic Business and Marketing*, SCI 484, pp. 125–136.
DOI: 10.1007/978-3-642-37932-1_9   © Springer-Verlag Berlin Heidelberg 2013

Based on TAM, this paper analyzes user behavior as to why people use video hosting websites. In particular, as extension of basic TAM, social influence and flow experience [12,13,14,15,16], which are often applied to Internet applications, are added. Moreover, the comforts of line speed, which is expected to be important in use of online services, and advertisement and charges that may prevent users from using online video services are also added. Based on the proposed hypothesis model, questionnaires are carried out and results of the questionnaires are analyzed to examine the proposed hypothesis.

2   Hypothesis Model for Video Hosting Services

2.1   Technology Acceptance Model

The Theory of Reasoned Action (TRA) shown as Fig. 1 is a widely studied model in social psychology to explain the behavior of an individual. According to TRA, an individualfs behavior is predicted by intentions and intentions are determined by his/her attitude and subjective norm concerning the behavior. Technology Acceptance Model (TAM), proposed by Davis, is an adaptation of TRA for modeling user acceptance of information systems. Fig. 2 shows TAM.

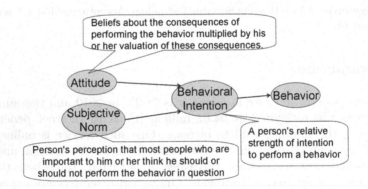

Fig. 1. Theory of Reasoned Action

In TAM consisted of six constructs, perceived usefulness (PU) and perceived ease of use (PE) are two primary determinants of information systemsf acceptance, and these beliefs were added. PU is defined as the degree to which a person believes that using a particular system will enhance his/her job performance. PE is defined as the degree to which using the technology will be free of effort. Both PU and PE are influenced by external variables such as information system qualities. PU is also influenced by PE. Attitude toward using (A) is jointly determined by PU and PE. A and PU predict behavioral intention to use (BI), and BI determines actual system use.

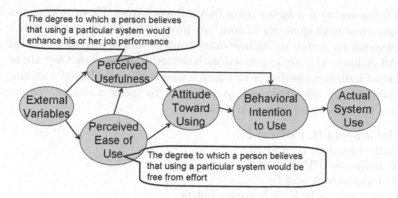

Fig. 2. Technology Acceptance Model

2.2   Extension for Video Hosting Services

Extending TAM explained above, a hypothesis model for online video service users is considered. As external variables, we examine a user's environment, including line speed, quantity of contents, advertisement and any charges that may prevent services from being used. In addition, "flow experience," which is often introduced to a system using the Internet, is applied. Flow is the holistic experience that people feel when they act with total involvement. Fig.3 shows an extended TAM as our hypothesis.

The factors in Fig. 3 are explained as follows. Comfortable Environment (C) is a factor that is concerned with usersf environment in accessing online video services and shows whether or not a user can watch video content comfortably.

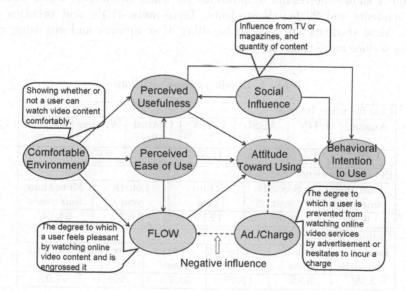

Fig. 3. Research hypothesis model

Social Influence(S) is a factor regarding influence from TV or magazines, and is also concerned with quantity of content. Flow (F) is the degree to which a user feels pleasant by watching online video content and is engrossed it. Influences from Ad./Charge (D) are expressed with dotted lines because they are negative. This factor indicates the degree to which a user is prevented from watching online video services by advertisement or hesitates to incur a charge. We supposed relations between these factors as follows.

- C influences PU, PE, and F.
- S influences PU, A, and BI.
- PE influences PU, F, and A.
- PU influences A and BI.
- When a user is in F, A becomes better.
- If A is improved, BI increases.
- F and A are weakened by D.

3   Questionnaire Survey

3.1   Survey Condition

To examine the above hypothesis model, a questionnaire was administered to students expected to use online video services. This questionnaire consists of 13 questions about user attributes (age, the years a user began using a PC, and so on) and 26 questions about 8 factors in the hypothesis model. The questionnaires asked 350 students to indicate their degree of agreement with the above questions based on a five-point scale: 5-strongly agree, 4-agree, 3-neutral, 2-disagree, and 1-strongly disagree.

Table 1 shows the results of questions for users' attributes. There were 330 male students and 20 female students. Teens were 41.1% and twenties were 57.4%. Most students are used to handling of computers and are using video hosting service more than one year.

Table 1. Results of users' attributes

Line for using Internet						
Analog	ISDN	ADSL	CATV	Optical fiber	Wireless	Others
1.4%	1.1%	27.1%	11.5%	42.7%	9.6%	6.6%

PC use experience				
Less than one year	Second year	Third year	Fourth year	More than four years
6.0%	8.9%	12.1%	8.6%	64.4%

Online video service experience					
Less than 3 months	A half year	One year	Two years	More than three years	Nothing
3.9%	5.8%	17.3%	30.3%	35.8%	7.0%

Table 2. Questionnaire results

	Strongly agree	Agree	Neutral	Disagree	Strongly disagree
C1	26.6%	25.7%	21.4%	13.7%	12.0%
C2	16.3%	18.6%	27.7%	18.9%	18.3%
C3	62.6%	19.1%	10.0%	4.0%	3.7%
C4	8.0%	22.3%	37.1%	18.6%	13.7%
PE1	40.3%	29.7%	18.6%	7.4%	3.4%
PE2	27.7%	34.6%	25.4%	9.1%	3.1%
PE3	26.6%	28.6%	29.1%	10.3%	5.1%
PE4	22.9%	29.7%	31.1%	11.1%	5.1%
S1	3.7%	8.9%	21.7%	16.6%	49.1%
S2	9.7%	15.7%	22.3%	12.9%	39.4%
S3	8.6%	12.6%	22.9%	14.6%	38.9%
F1	24.3%	26.0%	19.7%	14.0%	15.4%
F2	6.6%	12.3%	19.1%	15.4%	46.0%
F3	5.4%	8.3%	11.7%	15.7%	58.3%
PU1	13.1%	26.6%	37.7%	12.3%	9.7%
PU2	33.7%	30.3%	23.4%	4.9%	5.1%
PU3	29.4%	21.4%	27.7%	10.3%	11.1%
D1	18.6%	22.6%	23.4%	13.7%	20.9%
D2	8.9%	5.7%	26.6%	11.7%	46.9%
D3	14.0%	12.0%	29.4%	9.7%	33.1%
D4	22.6%	17.1%	31.7%	9.1%	18.6%
D5	49.4%	22.3%	18.9%	2.3%	6.6%
A1	11.7%	21.1%	38.6%	11.7%	16.3%
A2	16.6%	26.3%	39.4%	8.3%	8.9%
A3	30.9%	24.0%	28.9%	6.6%	9.1%
BI1	55.7%	21.4%	17.1%	2.9%	2.3%

3.2   Questionnaire Results

A part of the questionnaire results is shown in Table 2. The alphabet of Q_no in the table shows that the question is concerning the factor in the hypothesis model. Questions written as symbols in the table are as follows.

C1 Are you satisfied with the line using Internet?
C2 Are you satisfied with your PC spec?
C3 Can you use a PC whenever you want?
C4 Are you satisfied with the picture quality of contents provided by online video services?
PE1 Can you watch video contents with online video services easily?
PE2 Can you find desired video contents easily?
PE3 Have you used online video services easily from the first time?
PE4 Can you watch video contents with online video services comfortably?
S1 Do you access an online video services site introduced by magazines?
S2 Do you access an online video services site introduced by TV programs?

S3 Do you use sites that have many links for online video services sites?

F1 Have you ever been engrossed in watching video contents with online video services?

F2 Have you ignored the telephone or e-mail when watching video content?

F3 Have you forgotten to take a meal when watching video content?

PU1 Are you satisfied with the lineup of contents provided by online video services?

PU2 Are you satisfied that you can watch desired video content by using an online video service?

PU3 Are you satisfied that you can watch contents without waiting for rebroadcast of TV programs or sale of DVD version?

D1 If advertisements pop up to contents, do you lose interest in watching video contents?

D2 Do you restrict yourself to using an online video service because of worry about a charge?

D3 If content is provided illegally, do you stop watching it?

D4 If you are warned that watching site is unsafe, do you stop watching it?

D5 If online video services are provided for free, do you want to use them more?

A1 Do you feel exciting when you watch video contents with online video service sites?

A2 Do you feel good watching video contents with online video service sites?

A3 Do you like to download and watch video content through the Internet?

BI1 Will you use online video services in the future?

4   Behavior Analysis

4.1   Correlation Analysis

To examine relations between factors from the questionnaire results, correlations were analyzed. Correlations are tested with significance at the 0.01 level based on the t-distribution, and it is decided whether or not factors are correlated. Most results have correlations between the questions about the same factor. Namely, participants similarly answered questions about the same factor. The following is a test of the hypothesis.

Influence from C. All questions about F have no correlation with C. Table 3 lists correlation coefficients between C and PE or PU. Values written in boldface type mean they have correlations. In regard to PE, there are many correlations and coefficients are larger. However, concerning PU, there are only four correlations and coefficients are smaller. Therefore, it is predicted that C influences PE, but has no effect on F, and there is small influence from C on PU.

Influence from S. There is no correlation between S and BI. Table 4 shows correlation coefficients between S and PU or A. As for PU, about half of them are recognized as correlations, but coefficients are smaller. It is not clear that S influences PU. On the other hand, most relations between S and A have correlations, and S influences A.

Table 3. Correlation coefficients of Comfortable Environment

@	PE1	PE2	PE3	PE4	PU1	PU2	PU3
C1	0.21	0.13	0.21	0.33	0.14	0.05	0.03
C2	0.09	0.06	0.05	0.24	0.12	0.06	-0.02
C3	0.26	0.26	0.27	0.25	0.17	0.16	0.15
C4	0.22	0.18	0.10	0.22	0.33	0.11	0.10

Table 4. Correlation coefficients of Social Influence

@	PU1	PU2	PU3	A1	A2	A3
S1	0.05	0.02	0.08	0.14	0.15	0.16
S2	0.03	0.13	0.20	0.10	0.06	0.25
S3	0.15	0.22	0.21	0.25	0.17	0.22

Influence from PE. It was not judged that F and PU have a correlation. Table 5 shows correlation coefficients between PE and PU or A. As for correlation with PU, most relations are significant and it is predicted that there is influence. About half of the correlation with A is judged as significant but the correlation coefficients are smaller. It is not clear that there is strong influence.

Table 5. Correlation coefficients of Perceived Ease of Use

@	PU1	PU2	PU3	A1	A2	A3
PE1	0.31	0.33	0.24	0.18	0.18	0.25
PE2	0.31	0.25	0.19	0.17	0.11	0.19
PE3	0.27	0.22	0.21	0.01	0.08	0.12
PE4	0.34	0.22	0.11	0.10	0.16	0.20

Influence from PU. Table 6 shows correlation coefficients between PU and A or BI. From this table, many relations are judged to have correlations. Moreover, the coefficients are larger compared with other coefficients and it is expected that there is strong influence.

Relations between A and F or BI. As shown in Table 7, all relations between F and A are judged to have a correlation and it is expected that there is strong influence. According to correlation coefficients between A and BI, it is predicted that there is influence.

Influence from D. Table 8 shows the correlation coefficient between D and F or A. From this table, as for A, only five relations are judged to have significance. Concerning F, relations are partly judged to have a correlation, but the coefficients are small and it is expected that there is not strong influence. However, there is no correlation between D and other factors except F and A. Compared with the other factors, F and A are influenced by D.

Table 6. Correlation coefficients of Perceived Usefulness

@	A1	A2	A3	BI1
PU1	0.24	0.31	0.13	0.28
PU2	0.28	0.35	0.19	0.37
PU3	0.25	0.32	0.23	0.31

Table 7. Correlation coefficients between A and F or BI

@	A1	A2	A3
F1	0.52	0.44	0.21
F2	0.30	0.31	0.17
F3	0.27	0.22	0.16
BI1	0.30	0.35	0.29

Result Model. Summarizing the above examinations, Fig. 4 shows a result model. Strong influences are expressed as thick lines.

4.2   Analysis by Structural Equation Modeling

With the questionnaire results, our hypothesis model is verified using structural equation modeling (SEM)[17,18]. Before examination of our hypothesis, it is verified that user behavior of video hosting websites can apply the original TAM.

Result of SEM for Original TAM. Fig. 5 shows the analysis results of SEM with original TAM. In this figure, representation of error variables is omitted. Table 9 shows the estimated parameters between the latent variables. In this table, *** means 0.001 significant level. The following are the fit criteria of this model. The goodness of fit index (GFI) is 0.945, the comparative fit index (CFI) is 0.937, and the root mean square error of approximation (RMSEA) is 0.068. If the GFI and CFI values are 1.0, the model perfectly fits to the data. Empirically, a desirable value is over 0.95 or 0.9. The RMSEA value is desirable if it is less than 0.05: if RMSEA is over 0.1, the model is regarded as unfit. According to this results, this analyzed model has a better fitness and user behavior of online video hosting services can be explained by TAM. Total effects for BI from PE is

Table 8. Correlation coefficients of Ad./Charge

@	F1	F2	F3	A1	A2	A3
D1	0.15	0.18	0.18	0.15	0.20	0.10
D2	0.06	0.17	0.16	0.07	0.09	-0.02
D3	0.08	0.07	0.14	0.07	0.04	-0.11
D4	0.06	0.05	0.09	-0.04	-0.02	-0.06
D5	0.19	0.05	0.03	0.18	0.18	0.20

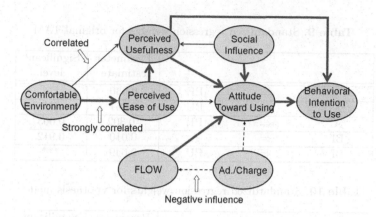

Fig. 4. Results of Correlation Analysis

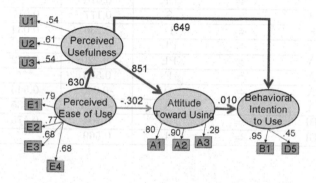

Fig. 5. Results of SEM for original TAM

0.411 and from PU is 0.657. Although both usefulness and ease of use have effect to behavioral intention, usefulness has more influence against other factors.

Result Model. Fig. 6 and Table 10 shows the analysis results. In this figure, representation of error variables and observed variables is omitted. As observed variables, the questionnaire is used for each belonging factor. The goodness of fit index (GFI) is 0.866, the comparative fit index (CFI) is 0.819, and the root mean square error of approximation (RMSEA) is 0.064. The GFI and CFI of this analyzed model are a little small and the whole fitness is not very good, but the model is not regarded as unfit. According to this result model, verified hypothetical relations are follows.

Table 9. Standardized regression weights for original TAM

			Parameter estimate	Significant level
PU	⇐	PE	0.630	***
A	⇐	PU	0.851	***
A	⇐	PE	-0.302	0.003
BI	⇐	A	0.010	0.912
BI	⇐	PU	0.649	***

Table 10. Standardized regression weights for ypothesis model

			Parameter estimate	Significant level
PE	⇐	C	0.469	***
PU	⇐	PE	0.616	***
PU	⇐	S	0.177	0.017
F	⇐	D	0.236	0.034
A	⇐	PU	0.720	***
A	⇐	PE	-0.220	0.022
A	⇐	F	0.533	***
A	⇐	D	-0.040	0.545
A	⇐	S	-0.053	0.435
BI	⇐	A	0.104	0.175
BI	⇐	PU	0.569	***

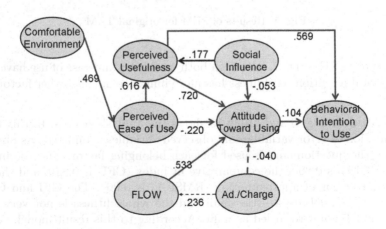

Fig. 6. Results of SEM

- C influences PE.
- S slightly influences PU.
- PE influences PU and A, although effect for A is negative.
- PU influences A and BI.
- When a user is in F, A becomes better.
- If A is improved, BI increases slightly.

Table 11 shows estimation of standardized total effects. Negative effects of advertisement or charge are not verified. It is guessed that students are using only free websites and do not feel advertisement annoying. The influence of hypothetical factors "flow" and "comfortable environment" are verified and these factors are important for online video hosting service.

Table 11. Total effects

	D	S	C	PE	PU	F	A
PE			0.469				
PU		0.177	0.289	0.616			
Flow	0.236						
A	0.086	0.075	0.105	0.224	0.720	0.533	
BI	0.009	0.108	0.175	0.374	0.644	0.055	0.104

5 Conclusion

Based on TAM, this paper analyzed user behavior for online video hosting services. As a hypothesis model that is an extended basic TAM, social influence and flow experience were added. Moreover, we also added comfort of line speed, which is expected to be important for using online services, along with advertisement and charges that may prevent users from using online video services. With questionnaire results, the hypothesis model was examined. Most relations were confirmed, but there is not a big influence from flow experience. It was found that advertisement and charges are not important factors in preventing use of online video services.

The questionnaire in this paper was given only to students, but it is not enough. Therefore, we should improve the questionnaire and examine wider target.

References

1. Ministry of Internal Affairs and Communications, Japan: Information and Communications in Japan (2011 WHITE PAPER),
   http://www.johotsusintokei.soumu.go.jp/whitepaper/
2. Watanabe, K., Iwasaki, K.: Factors Affecting Consumer Decisions about Purchases at Online Shops and Stores. In: Proceedings of IEEE CEC/EEE 2007, pp. 80–87 (2007)

3. Hiramatsu, A., Yamasaki, T., Nose, K.: Survey of Consumers' Decision Making Process for Online Music Service. In: Proceedings of ICETE 2008 International Conference on e-Business (ICE-B 2008), pp. 229–234 (2008)
4. Hiramatsu, A., Yamasaki, T., Nose, K.: Decision Making Model for Online Music Service Users, Towards Sustainable Society on Ubiquitous Networks. In: Proc. of The 8th IFIP Conference on e-Business, e-Service, and e-Society (I3E 2008), pp. 15–25 (2008)
5. Davis, F.D.: Perceived Usefulness, Perceived Ease of Use, and User Acceptance of Information Technology. MIS Quarterly 13(3), 319–340 (1989)
6. Hsu, C.L., Lu, H.P.: Why do people play on-line games? An extended TAM with social influences and flow experience. Information & Management 41, 853–868 (2004)
7. Shih, H.P.: An empirical study on predicting user acceptance of e-shopping on the Web. Information & Management 41, 351–368 (2004)
8. Liu, C.: Modeling Consumer Adoption of The Internet as a Shopping Medium. Cambria Press (2007)
9. Jung, Y., Perez-Mira, B., Wiley-Patton, S.: Consumer adoption of mobile TV: Examining psychological flow and media content. Computers in Human Behavior 25(1), 123–129 (2009)
10. Liao, C.H., Tsou, C.W.: User acceptance of computer-mediated communication: The SkypeOut case. Expert Systems with Applications 36(3), Part 1, 4595–4603 (2009)
11. Johari, S.S.M.: The Acceptance Model of Mobile Learning. Lambert Academic Publishing (2012)
12. Novak, T.P., Hoffman, D.L., Duhachek, A.: The Influence of Goal-Directed and Experiential Activities on Online Flow Experiences. Journal of Consumer Psychology 13(1-2), 3–16 (2003)
13. Pace, S.: A grounded theory of the flow experiences of Web users. International Journal of Human-Computer Studies 60(3), 327–363 (2004)
14. Pilke, E.M.: Flow experiences in information technology use. International Journal of Human-Computer Studies 61(3), 347–357 (2004)
15. Thatcher, A., Wretschko, G., Fridjhon, P.: Online flow experiences, problematic Internet use and Internet procrastination. Computers in Human Behavior 24(5), 2236–2254 (2008)
16. Koufaris, M.: Applying the Technology Acceptance Model and Flow Theory to Online Consumer Behavior. Information Systems Research 13(2), 205–223 (2002)
17. Schumacker, R.E., Lomax, R.G.: A Beginnerf's Guide to Structural Equation Modeling. Lawrence Erlbaum Associates (2004)
18. Kline, R.B.: Principles and Practice of Structural Equation Modeling. The Guilford Press (2005)

Social Media Networker: A New Profile
for a New Market

Aniko Petakne Balogh[1], Kerstin V. Siakas[2], Sonja Koinig[3], Damjan Ekert[3],
Darragh Coakley[4], Ricardo Colomo-Palacios[5], and Vassilis Kostoglou[2]

[1] Denis Gabor College, Hungary
[2] Alexander Technological Educational Institute of Thessaloniki, Greece
[3] ISCN, Austria
[4] Cork Institute of Technology, Ireland
[5] Universidad Carlos III de Madrid, Spain

Abstract. The emergence of Internet-based social media has enabled social networks for online collaboration, interaction, communication and knowledge sharing between users who have common interests, needs or goals. The social media phenomenon has changed the way people communicate with each other and the way organizations and governments interact with different stakeholders. Consumers are increasingly utilizing Web 2.0 platforms, such as content sharing sites, blogs, social networks, and wikis in order to create, modify, share, and discuss online content. Communities are built from the bottom-up by people who capitalize on the behavior and knowledge of others. Barriers of distance and time are overcome by transparency created by the wisdom of the users. The use of social media has had a significant impact on public relations and the reputation of organizations, not to mention marketing, sales and sustainability. Yet, many organizations eschew or ignore this type of media, mainly because they do not understand the advantages of its use, the various forms it can take, and how to engage in it. In order to address this lack of skills and to support organizations in building competences in the field of social media a lifelong learning course is currently under development within the frame of the European Certification and Qualification Association (ECQA). The ECQA Social Media Networker (SIMS) is a two year long project which started in October 2011 with funding from the EU Lifelong Learning Programme (LLP). The project aims to develop a new skill set and a job role qualification study program, where competencies in social media are customized for the European industry into an online study program complemented with an online training, examination and certification schema for the Social Media Networker job role. A pilot training scheme will take place in the participating organizations/member states (Austria, Greece, Hungary, Ireland and Spain) and the study programme will be refined and improved based on systematic feedback. This chapter describes the joint effort of five organizations in five countries to define and develop the professional role of a Social Media Networker. After developing and testing the modular training material, the profession will be launched to the market. Future work will concentrate on the sustainability of the profession by the creation of a job role committee that will be responsible for updating the training material, as well as for the dissemination and exploitation of the results.

T. Matsuo & R. Colomo-Palacios (Eds.): *Electronic Business and Marketing*, SCI 484, pp. 137–146.
DOI: 10.1007/978-3-642-37932-1_10   © Springer-Verlag Berlin Heidelberg 2013

Keywords: Social media, ECQA, profession, networker, Social Web, Web 2.0.

1   Introduction

The Internet has changed the way people communicate, but also the way organizations interact with customers, employees and partners. In other words, Internet has profoundly changed the human experience [7]. The emergence of Internet-based social media has made it possible for one person or organization to communicate with hundreds or even thousands of other people about products and the companies that provide them [19]. The current trend toward Social Media can therefore be seen as a realization of the Internet's roots, since it re-transforms the World Wide Web to what it was initially created for: a platform to facilitate information exchange between users [17]. Social media encompasses a wide range of online tools including blogs, company-sponsored discussion boards and chat rooms, consumer-to-consumer e-mail, consumer product or service ratings websites and forums, Internet discussion boards and forums, moblogs (sites containing digital audio, images, movies, or photographs), and social networking websites, but to name a few [19]. As a consequence of its usefulness Social Media has been pointed out as useful for human fields, including public relations [8][12], crisis management [1][32][34], eGovernment [3][4][6][9][10], marketing [2][30][31], Education [11][14][15][25], Tourism [21][26][28][33], Sports [22] and medicine [13][16] citing the most relevant and recent research works.

However, this social media strategy requires the participation of forward-thinking individuals, who must be socially and technically knowledgeable and informed in order to exploit these new ways of social networking. All stakeholders in an organization can benefit from understanding social media networking principles and their implementation. However, this is not an easy task. To correctly facilitate these tools, one must learn how to leverage them in order to improve marketing, research, communications, customer support, brand reputation, competitive intelligence, product development, collaboration and knowledge capture. Although there are recent and relevant works designed to illustrate some of the skills and activities of social media professionals [27][29], to the best of authors' knowledge, there is currently no initiative to characterize the competencies of this new type of professional. This chapter is a joint effort of organizations from five different European countries to define this professional role.

The project history is as follows: in 2011 five organizations decided to submit the project proposal titled "ECQA Certified Social Media Networker Skills". The project partners are the following institutions: Universidad Carlos III de Madrid (Spain) I.S.C.N. GesmbH (Austria), Dennis Gabor College, DGC (Hungary), Alexander Technological Educational Institute of Thessaloniki, ATEI (Greece), and the Cork Institute of Technology, DEIS Department of Education Development (Ireland). All partners had considerable previous experience within their field and their cooperation added value and represented a synergy of expertise in different areas. This chapter describes the outcomes of the project in terms of learning content and objectives.

The remainder of the chapter is structured as follows: The next section is devoted to an explanation depict the European Certification and Qualification Association (ECQA) along with its custom online teaching platform. The subsequent section illustrates the Social Media Networker Project, a project funded by Leonardo Da Vinci European Program to design this new professional role. Following this, the initiative is discussed and, finally, section 5 concludes the chapter and outlines future research plans.

2 The European Certification and Qualification Association

The European Certification and Qualification Association (ECQA) is a non-profit association connecting different organizations (companies, university, institutions etc.) and thousands of professionals world-wide via training and certification for a wide range of job roles - currently the ECQA offers training and certification for 30 job roles. New job roles are also constantly being developed. The ECQA provides its services in more than 24 countries through more than 60 members (consisting of training organizations, trainers, exam organizations etc.).

2.1 Background

The ECQA is the result of a series of different projects co-funded by the European Commission under the Life Long Learning programme:

The EQN (European Quality Network, 2005 - 2007) established a network of different members (previous Life Long Learning Leonardo projects, European networks, chambers of commerce, trainings organizations and institutes), who have developed qualification programmes and training courses. These different groups jointly analysed key quality indicators and success factors to achieve high quality services and a continuous innovation process. The result was a set of quality rules for job role based qualifications related to skills architecture, course syllabus, test questions and certification rules. These quality criteria set the framework for the foundation of the European Certification and Qualification Association.

The EU Cert Campus project (2008-2009) established an online training campus with online services, which were implemented in the ECQA infrastructure:

- Skill set browsing
- Central registration service
- Multilanguage support
- Online trainings through a learning management system.
- Self maintenance of the skill sets and self test / examination pool

With the support of the ECQA services and the EU Cert Campus more than 200 participants attended online trainings from their work place and/or home.

The dEUCert Project (Dissemination of EU Certificates) aimed to disseminate the ECQA framework Europe-wide by building ECQA regional contact centers represented by ECQA ambassadors. ECQA Ambassadors are persons with an outstanding success record on an international level, which are selected because of their substantial contribution to the ECQA and to their support to create a European vision of collaboration based on a joint educational and certification strategy on the market.

2.2 Quality Criteria

The ECQA ensures that the same core knowledge is presented to all training participants by defining so called "skill sets" (representing the content, subject matter, learning objectives, etc.). All exam participants are tested according to the same requirements (quality criteria) contained in the skill sets. These quality criteria apply to the following types of service providers:

- Trainers – performing the ECQA certified trainings
- Training organizations – organizing and offering trainings
- Exam organizations – organizing and supervising exams
- Certification organizations – national certification bodies and organization issuing certificates in the name of the ECQA

The content of the job roles/trainings is defined by experts from industry and research by combining profound knowledge with research results and best practices. These experts collaborate in ECQA groups referred to as Job Role Committees. The main responsibilities of the Job Role Committees are:

- To maintain and regularly review the actuality of the skill set;
- To prepare and maintain the examination question pool (selected members of the Job Role Committee operate as Exam Committee);
- To review and approve new training material, trainers and training organizations.

The aim is to ensure the same quality level of training and certification in all participating countries.

2.3 Skill Set Definition

The skill set (syllabus) for the Social Media Networker provides the basis for the development of training and examination. The skill set is based on the skills definition proposed by the DTI (Department of Trade and Industry) in the UK for the NVQ (National Vocational Qualification).

   The defined skill set follows the ECQA compliant skills definition standards and describes the target profile's competences in terms of skill units, which are defined by skill elements. Each skill element consists of a number of performance criteria (learning outcomes) describing the minimum level of performance, which a training participant has to demonstrate in order to be certified as an "ECQA certified Social Media Networker".

   Each skill set consists of the following items:

- **Domain:** An occupational category (e.g. Domain "Social Media Management")
- **Job Role:** A job description that covers part of the domain knowledge (e.g. someone in the job role of a "Social Media Networker")

- **Unit:** A list of certain activities that have to be carried out in the workplace. It is the top-level skill in the qualification standard hierarchy. Each unit consists of a number of elements.
- **Learning Element:** A description of one distinct aspect of the work performed by a worker, either a specific task that the worker has to do or a specific way of working. Each element consists of a number of performance criteria.
- **Performance criteria:** Description of the minimum level of performance, which a training participant has to demonstrate in order to be assessed as someone qualified for the respective job role – e.g.: "ECQ certified Social Media Networker".
- **Optional Level of cognition:** For each performance criteria there is an intended level of cognition. At the same time this describes the complexity level of the exam questions for each performance criteria (according to Bloom's Taxonomy).

2.4   ECQA Online Services

Through the support of the EU Cert Campus project, the partnership developed a skill assessment and online learning portfolio portal supporting procedures for account management, browsing of skill set and self assessment. In addition, the participant can prove required skills by:

- uploading reference materials, which are evaluated by external assessors in a formal assessment and/or by:
- attending an online examination.

The ECQA learning and examination procedure consists of the following steps:

1. The participant visits the ECQA webpage (www.ecqa.org) and browses through the different skill sets.
2. The participant registers at the ECQA learning portal, performs a self assessment examination and receives a skill gap analysis identifying which skills they are proficient in and which skills are not proficient in.
3. Based on this analysis the participant attends an online or on-site training.
4. After the training the participant attends an online multiple-choice examination and after successfully passing this, a certificate is issued by the ECQA. The exam questions are randomly selected from a larger pool of questions resulting in a personalized exam for every participant.

As a consequence of the importance of the ECQA initiatives, there are many research papers in which many of the professions that are already included in the ECQA portfolio have been depicted e.g. [5][18][20][23][24].

3   The Social Media Networker Project

The innovation transfer project ECQA Certified Social Media Networker Skills (SIMS) deals with the professional qualification and certification of the job role of Social Media Networker. The aim of this project is the transfer of the Social Media

Networker Skill to industry. This will provide a technological and methodological strategy for online learning assessment, recognition, facilitation and qualification for this new job role. The project includes:

- Development of an online accredited programme of study for the job role of a social media networker.
- Piloting of the programme in all participating member states of the project and refinement of the programme based on systematic feedback.
- Dissemination of the project results via a range of channels including a major conference and special issues in journals.
- Ensuring the sustainability of the project and its outputs through the European Certificates Association and other bodies.

To this aim, this project will establish a skill set and certification criteria based on the European-wide accepted scheme of the ECQA. It will involve industrialists on an international level, and will implement the results in the respective partner institutions.

The project partners arranged the project kick off meeting in Madrid in December 2011. Here they decided on the initial skill set titles of the Social Media Networker course and distributed the task of writing 2-3 skill card elements each and the corresponding training material based on the skill card based on literature review and self-initiated research. For quality management purposes two other partners reviewed the skill cards independently.

The structure of the skill cards (and consequently the training material and the course) complies with the ECQA modularity rules for professions. The content of a profession should be divided into the:

- Domain: The domain is the name of the profession (e.g. ECQA Social Media Networker Manager)
- Unit: The content of the training is grouped into logical subject matter topics (units). The number of units should be 3 to 7 per domain
- Elements: In order to provide a better structuring of the units, the units are divided into elements. The suggested number of elements is 3 to 7 per unit.
- Performance criteria (PC): They are the criteria set for the minimum level of knowledge and performance required for a participant to effectively function within the given job role. Performance criteria are defined for each element and the suggested number of PCs is 3 to 6 per element.

The content of a Skills card is used within the ECQA for several purposes:

- for the description of the profession in dissemination materials and on the ECQA web page,
- for the structuring the ECQA exam portal,
- for the design of certificates for participants, etc.

For each profession a short identifier consisting of three characters should be selected. The identifier is used in the exam portal, on certificates, etc. (e.g. SRM for Social

Responsibility Manager, MAN for EU Project Manager, etc.). Availability of the profession code is checked by the ECQA.

For the job role of a social media networker, the topics of the skill cards are the following:

Unit 1:

- E1: Introduction to Social Media
- E2: Social Media Technologies

Unit 2:

- E1: Social and Business Networks (Facebook, Google+, LinkedIn, XING, asmallWorld, MySpace...)E2: Blogging, Microblogging (Twitter, Bloger, Wordpress, Jaiku, Foursquare)
- E3: Content Sharing, Recommendation and Collaboration (Youtube, Flickr, Picassa, Podcast, Goodle Docs, Wiki, Dropbox, Slideshare , LastFM, Genius, Pandora, Digg, Amazon, Snooth ...) + Education

Unit 3:

- E1: Planning, Implementing and Monitoring Communication (Trends, Analytics, Integration to the whole communication plan ....)
- E2: Enterprise 2.0 (Knowledge Management , HRM, Co-creation and user generated content, Training)
- E3: Marketing and CRM (Social Media Marketing ,CRM and Supply/Chain and Social Media)

Unit 4:

- E1: Culture of sharing and Online Reputation Handling (Management), Use of Language, Branding, Netiquette
- E2: Legal and Financial aspects of social media (copyright content and Culture of sharing)
- E3: Information overload (Semantic Web, Data Mining, Natural Language Processing)

4   Discussion

There is no doubt that research and industry have access to, via social media tools, a new environment of communication and collaboration. Thus, there is a need to facilitate the relevant stakeholders (owners, shareholders, suppliers, employees, researchers, customers, and the broader community) in their successful adaptation to this environment. Social networking is currently the spearhead for a significant extension in this field, connecting researchers, developers, customers, supply chains, or creating Enterprise 2.0 Knowledge Management (KM) policies based on social networking as a strategy will be a decisive factor for future success for many businesses.

The Social Media Networker profession arose from the essential need of the organizations to get involved both actively and efficiently in social networking. The main aim of this profession is to exploit these new ways of social networking and their technologies and principles for communication and collaboration, as well as detailing how organizations can profit from their implementation. Production of reliable and up to date educational material, provision of structured and uniform training in the countries of the consortium members, and setting up online exams will be the initial steps. Afterwards, as the importance and use of social media is anticipated to increase dramatically, training for this job role will need to be disseminated and exploited, so that it becomes a useful mainstream tool in the lifelong learning processes. Furthermore, as the development of relevant technologies is so rapid (Web 3.0., semantic web concentrating on data meaning, personalization, intelligent search, and not only) training material will need to be updated frequently. This issue will be of vital importance for the desirable successful continuation of this profession and the increase in the number of relevant experts.

5   Conclusions and Future Work

This chapter focused on a new promising profession - the Social Media Networker. After the presentation of a relevant literature review, the concept and activities of the European Certification and Qualification Association were described. The next sections were devoted to the illustration of the Social Media Networker Project covering all its main issues: establishment, concept and initiatives, terminology, as well as training material contents.

The project consortium members believe the future trained professionals will enjoy high employability and promising careers within the EU member-states, and beyond. Relevant future work will initially concentrate on the sustainability of the Social Media Networker profession. After launching the profession on the market a job role committee will be created by the current consortium members. The job role committee will be responsible for updating the training material with new technological and social trends. Additionally, as it is obvious that the use of social networking will be increased exponentially in the forthcoming years, the creation of even more relevant new professions suggests promising dynamics in this field. These new professions should focus on social media technology and marketing issues. The consortium of the present project plans to focus on such research activities in coming years.

Acknowledgements. This work is supported by the European Commission (Programme LifeLong Learning - action Leonardo da Vinci-Transfer of Innovation); 2011-1-ES1_LEO05-35930.

References

1. Alfonso, G.-H., Suzanne, S.: Crisis Communications Management on the Web: How Internet-Based Technologies are Changing the Way Public Relations Professionals Handle Business Crises. Journal of Contingencies and Crisis Management 16(3), 143–153 (2008), doi:10.1111/j.1468-5973.2008.00543.x
2. Berthon, P.R., Pitt, L.F., Plangger, K., Shapiro, D.: Marketing meets Web 2.0, social media, and creative consumers: Implications for international marketing strategy. Business Horizons 55(3), 261–271 (2012), doi:10.1016/j.bushor.2012.01.007
3. Bertot, J.C., Jaeger, P.T., Grimes, J.M.: Using ICTs to create a culture of transparency: E-government and social media as openness and anti-corruption tools for societies. Government Information Quarterly 27(3), 264–271 (2010), doi:10.1016/j.giq.2010.03.001
4. Bertot, J.C., Jaeger, P.T., Munson, S., Glaisyer, T.: Social Media Technology and Government Transparency. Computer 43(11), 53–59 (2010), doi:10.1109/MC.2010.325
5. Biró, M., Messnarz, R.: SPI experiences and innovation for Global Software Development. Software Process: Improvement and Practice 14(5), 243–245 (2009), doi:10.1002/spip.432
6. Bonsón, E., Torres, L., Royo, S., Flores, F.: Local e-government 2.0: Social media and corporate transparency in municipalities. Government Information Quarterly 29(2), 123–132 (2012), doi:10.1016/j.giq.2011.10.001
7. Correa, T., Hinsley, A.W., de Zúñiga, H.G.: Who interacts on the Web?: The intersection of users' personality and social media use. Computers in Human Behavior 26(2), 247–253 (2010), doi:10.1016/j.chb.2009.09.003
8. Curtis, L., Edwards, C., Fraser, K.L., Gudelsky, S., Holmquist, J., Thornton, K., Sweetser, K.D.: Adoption of social media for public relations by nonprofit organizations. Public Relations Review 36(1), 90–92 (2010), doi:10.1016/j.pubrev.2009.10.003
9. Chen, H.: AI, E-government, and Politics 2.0. IEEE Intelligent Systems 24(5), 64–86 (2009), doi:10.1109/MIS.2009.91
10. Chun, S.A., Shulman, S., Sandoval, R., Hovy, E.: Government 2.0: Making connections between citizens, data and government. Information Polity 15(1), 1–9 (2010), doi:10.3233/IP-2010-0205
11. Ebner, M., Lienhardt, C., Rohs, M., Meyer, I.: Microblogs in Higher Education – A chance to facilitate informal and process-oriented learning? Computers & Education 55(1), 92–100 (2010), doi:10.1016/j.compedu.2009.12.006
12. Eyrich, N., Padman, M.L., Sweetser, K.D.: PR practitioners' use of social media tools and communication technology. Public Relations Review 34(4), 412–414 (2008), doi:10.1016/j.pubrev.2008.09.010
13. Greysen, S., Kind, T., Chretien, K.: Online Professionalism and the Mirror of Social Media. Journal of General Internal Medicine 25(11), 1227–1229 (2010), doi:10.1007/s11606-010-1447-1
14. Grosseck, G.: To use or not to use web 2.0 in higher education? Procedia - Social and Behavioral Sciences 1(1), 478–482 (2009), doi:10.1016/j.sbspro.2009.01.087
15. Hemmi, A., Bayne, S., Land, R.: The appropriation and repurposing of social technologies in higher education. Journal of Computer Assisted Learning 25(1), 19–30 (2009), doi:10.1111/j.1365-2729.2008.00306.x
16. Jorgensen, G.: Social media basics for orthodontists. American Journal of Orthodontics and Dentofacial Orthopedics 141(4), 510–515 (2012), doi:10.1016/j.ajodo.2012.01.002
17. Kaplan, A.M., Haenlein, M.: Users of the world, unite! The challenges and opportunities of Social Media. Business Horizons 53(1), 59–68 (2010), doi:10.1016/j.bushor.2009.09.003

18. Korsaa, M., Biro, M., Messnarz, R., Johansen, J., Vohwinkel, D., Nevalainen, R., Schweigert, T.: The SPI manifesto and the ECQA SPI manager certification scheme. Journal of Software Maintenance and Evolution: Research and Practice, n/a–n/a (2010), doi:10.1002/smr.502
19. Mangold, W.G., Faulds, D.J.: Social media: The new hybrid element of the promotion mix. Business Horizons 52(4), 357–365 (2009), doi:10.1016/j.bushor.2009.03.002
20. Nevalainen, R., Schweigert, T.: A European scheme for software process improvement manager training and certification. Journal of Software Maintenance and Evolution: Research and Practice 22(4), 269–277 (2010), doi:10.1002/spip.438
21. Parra-López, E., Bulchand-Gidumal, J., Gutiérrez-Taño, D., Díaz-Armas, R.: Intentions to use social media in organizing and taking vacation trips. Computers in Human Behavior 27(2), 640–654 (2011), doi:10.1016/j.chb.2010.05.022
22. Pfahl, M.E., Kreutzer, A., Maleski, M., Lillibridge, J., Ryznar, J.: If you build it, will they come?: A case study of digital spaces and brand in the National Basketball Association. Sport Management Review, doi:10.1016/j.smr.2012.03.004
23. Riel, A., Tichkiewitch, S., Messnarz, R.: Qualification and certification for the competitive edge in Integrated Design. CIRP Journal of Manufacturing Science and Technology 2(4), 279–289 (2010), doi:10.1016/j.cirpj.2010.04.005
24. Riel, A., Tichkiewitch, S., Messnarz, R.: Integrated engineering skills for improving the system competence level. Software Process: Improvement and Practice 14(6), 325–335 (2009), doi:10.1002/spip.424
25. Roblyer, M.D., McDaniel, M., Webb, M., Herman, J., Witty, J.V.: Findings on Facebook in higher education: A comparison of college faculty and student uses and perceptions of social networking sites. The Internet and Higher Education 13(3), 134–140 (2010), doi:10.1016/j.iheduc.2010.03.002
26. Schmallegger, D., Carson, D.: Blogs in Tourism: Changing Approaches to Information Exchange. Journal of Vacation Marketing 14(2), 99–110 (2008), doi:10.1177/1356766707087519
27. Siculiano, V.: The Direction of the Media Profession. International Journal on Media Management 13(3), 205–209 (2011), doi:10.1080/14241277.2011.576966
28. Thevenot, G.: Blogging as a Social Media. Tourism and Hospitality Research 7(3-4), 287–289 (2007), doi:10.1057/palgrave.thr.6050062
29. Vuori, M.: Exploring uses of social media in a global corporation. Journal of Systems and Information Technology 14(2), 155–170 (2012), doi:10.1108/13287261211232171
30. Wang, X., Yu, C., Wei, Y.: Social Media Peer Communication and Impacts on Purchase Intentions: A Consumer Socialization Framework. Journal of Interactive Marketing, doi:10.1016/j.intmar.2011.11.004
31. Weinberg, B.D., Pehlivan, E.: Social spending: Managing the social media mix. Business Horizons 54(3), 275–282 (2011), doi:10.1016/j.bushor.2011.01.008
32. White, C., Plotnick, L., Kushma, J., Hiltz, S.R., Turoff, M.: An online social network for emergency management. International Journal of Emergency Management 6(3), 369–382 (2009), doi:10.1504/IJEM.2009.031572
33. Xiang, Z., Gretzel, U.: Role of social media in online travel information search. Tourism Management 31(2), 179–188 (2010), doi:10.1016/j.tourman.2009.02.016
34. Yates, D., Paquette, S.: Emergency knowledge management and social media technologies: A case study of the 2010 Haitian earthquake. International Journal of Information Management 31(1), 6–13 (2011), doi:10.1016/j.ijinfomgt.2010.10.001

Context-Aware Advertising in Pervasive Computing Environment

Takuya Maekawa

Graduate School of Information Science and Technology, Osaka University,
2-1 Yamadaoka, Suita, Osaka, 565-0871, Japan
maekawa@ist.osaka-u.ac.jp
http://www-komo.ise.eng.osaka-u.ac.jp/~maekawa/

Abstract. Due to the recent advances in sensing and wireless communication technologies, we have been able to capture and understand real-world phenomena by employing tiny ubiquitous sensors such as accelerometers, thermometers, and RFID tags installed in daily environments. For example, by attaching ubiquitous sensors to various indoor objects and furniture, we can observe their use and phenomena that occur around them. Real world context information obtained from the ubiquitous sensors has triggered a wide range of applications in, for example, context-aware systems, lifelogging, and monitoring. This article describes a new type of context-aware advertising that employs context information obtained by ubiquitous sensors. We also introduce examples of context-aware advertising on our lifelogging and recommender systems.

Keywords: Advertising, pervasive computing, context-awareness, sensors.

1 Introduction

Recent advances in Internet and information retrieval technologies has enabled companies to perform targeted advertising. That is, an advertisement is provided according to a user's profile and search keywords. Also, the recent proliferation of positioning technologies including Global Positioning System (GPS) has triggered location-based advertising. That is, an advertisement is provided according to a user's location. For example, an advertisement related to a nearby shopping mall is provided to the user. On the other hand, due to the recent advances in sensing and wireless communication technologies, many studies have tried to capture and understand real-world phenomena by employing tiny ubiquitous sensors such as accelerometers, thermometers, and RFID tags. For example, by attaching tiny ubiquitous sensors to various indoor objects and furniture, we can observe their use and phenomena that occur around them. Context information obtained from the ubiquitous sensors has triggered a wide range of applications in, for example, context-aware systems, lifelogging, and monitoring. This article describes a new type of context-aware advertising that employs context information obtained by ubiquitous sensors.

T. Matsuo & R. Colomo-Palacios (Eds.): *Electronic Business and Marketing*, SCI 484, pp. 147–165.
DOI: 10.1007/978-3-642-37932-1_11　　　　　　© Springer-Verlag Berlin Heidelberg 2013

We assume that a context-aware application that employs ubiquitous sensors (e.g., lifelogging and recommender systems) is provided to a user. The application provides an advertisement that relates to the user's current situation in addition to its service. (The advertisement is included in the service.) Existing sensing technologies have provided a location-aware advertisement that relates to the user's current location by employing GPS sensors. By employing ubiquitous sensors, we can provide an effective advertisement that relates to the user's current activity and detailed situation. For example, when a user is brewing tea, an advertisement related to seasonal green tea products can be provided to the user. Also, when a user is cooking, an advertisement related to a recipe book can be provided to the user. In this article, we introduce examples of context-aware advertising that employs ubiquitous sensors on lifelogging and recommender systems.

In the rest of this article, we first introduce work related to existing advertising technologies and recent sensing technologies. Then we describe context-aware advertising that employs ubiquitous sensors. After that, we introduce examples of context-aware advertising on lifelogging and recommender systems.

2   Related Work

2.1   Advertising Technologies

Advertising is the major source of revenue for many TV stations, radio stations, and web sites. Existing advertising techniques are mainly categorized into the following groups.

- **Untargeted advertising:** The majority of advertisements on the web pages, TV programs, and radio programs are categorized into this advertising technique. This technique basically does not utilize information about a user. (Several advertisements target an expected user. For example, a sport manufacturer may place an advertisement related to sports products on a web page related to sports.)
- **Content-based advertising:** This technique utilizes the text of a web page that a user is browsing or search keywords that the user inputs. An advertisement is automatically associated with the textual information, and then the advertisement is provided to the user. Many studies have tried to associate an advertisement with the textual information. For example, in [17], the authors attempt to adapt online advertisement to a user's short-term interests by employing search keywords provided by the user to search engines. In [40], the authors associate an advertisement with web page text by expanding keywords associated with the advertisement and the web page text. In [49], the authors attempt to find appropriate keywords from web pages by employing supervised machine learning approaches. The found keywords are used to find appropriate advertisements. In this content-based advertising technique, the textual information (obtained from the web page or search keywords) and keywords associated with the advertisement are represented

as vectors of word weights. We employ some similarity measures such as cosine similarity to find an appropriate advertisement vector [41,2].

- **Profile-based advertising:** This technique utilizes a user's profile such as web browsing histories, buying histories, and keywords input by the user in advance. (The keywords show preference of the user.) SMMART is a mobile terminal-based advertising framework that provides advertisements by employing user input keywords [15]. MALCR is also a mobile terminal-based advertising framework that extracts a user's profile from her browsing behavior [50]. Many studies and web sites such as Amazon.com employ collaborative filtering techniques [19,11]. In this collaborative filtering techniques, the similarity between users is computed in advance by employing their profiles. The underlying assumption of the collaborative filtering is that if user A is similar to user B, user A may like an item (e.g., product) that user B likes. Based on the assumption, advertisements are provided to users.
- **Context-aware advertising:** This technique utilizes a user's context information to provide an advertisement to the user. Many studies employ location information as context information [1,7]. Above mentioned mobile terminal-based advertising studies (e.g., SMMART and MALCR [15,50]) employ the user's (mobile terminal's) location information in addition to the user's profile. In SMMART, for example, a user receives an advertisement related to books when she is at a book store.

2.2   Sensing Technologies

Location Sensing. Recently mobile phones have been equipped with various kinds of sensors, and we can benefit from the context information that they provide. In particular, latitude and longitude information obtained from a GPS sensor is widely used in various applications. However, GPS-based positioning can not be deployed for indoor use because line-of-sight transmission between the GPS sensor and satellites is not possible in indoor environments. To determine positions indoors, WiFi modules incorporated in mobile phones are usually used. Many services and studies employ the fingerprint-based WiFi positioning method [3,38]. The fingerprinting method refers to techniques that match the fingerprint of location-dependent characteristics, i.e., the received signal strength from a WiFi access point (AP). The fingerprints obtained at different locations using a *war-driving* technique [16] are stored in a database associated with their coordinates, and compared with the current fingerprints to determine the user's current coordinates.

Activity Sensing. Daily activity sensing/recognition is one of the most important tasks in pervasive computing applications because it has a wide range of uses in, for example, supporting the care of the elderly, lifelogging, and home automation [35,8].

Two main approaches are used for activity recognition studies: environment augmentation and wearable sensing. The environment augmentation approach attempts to recognize users' activities by using sensors embedded in indoor environments. In the computer vision community, activity recognition tasks are accomplished by using cameras installed in a given environment. For example, hand washing and operating medical appliances can be recognized by domain specific solutions [34,42]. However, the task has become dominated by various types of embedded small sensors. Recently, many researchers in the field of ubiquitous computing have tried to recognize activities based on dense object usage sensors such as RFID tags and switch sensors installed in indoor environments [36,45,47]. With this approach, many studies recognize activities of daily living such as using the toilet, making coffee, washing dishes, and taking medicine by using object usage sensors that are embedded in or attached to such daily use indoor objects and appliances as toilets, coffee makers, sinks, and cups. In [48], the authors recognize kitchen activities by using RFID tags attached to kitchen objects and a camera that overlooks the kitchen counter. An RFID reader bracelet worn on a user's wrist and the camera detect the use of the objects. In [24], we recognize the use of portable electrical devices by using battery-shaped sensor nodes that can be inserted into the portable electrical devices.

The wearable sensing approach tries to recognize a user's activities by employing such sensors as body-worn accelerometers and microphones to capture characteristic repetitive motions, postures, and sounds of activities [21,22,4,39,5,18,26]. Using these types of wearable sensors, sensing studies have successfully recognized such activities as walking, bicycling, brushing teeth, speaking and laughing, and workshop activities such as sawing and drilling that have characteristic motions and/or sounds. In [6], the authors use a camera and a microphone attached to a chest strap to detect location related events such as entering an office, kitchen, or courtyard. In [27], we attempt to recognize daily activities by using a wrist-worn sensor device with an accelerometer, a microphone, and a camera. The systems proposed in [23,25] recognize the use of portable electrical devices held by a user by employing several magnetic sensors attached to her hands. The systems capture magnetic fields emitted by magnetic components such as coils and permanent magnets in portable electrical devices and identifies which electrical device the user is using.

3   Activity-Based Advertising

As mentioned above, we can capture various real world events with recent sensing technologies. That is, we can provide an advertisement according to such real world events. In addition to location information, we can employ the following real world information.

- **Daily activity:** We can provide an effective advertisement that relates to a user's current activity. For example, when a user is brewing tea, we can provide an advertisement related to seasonal green tea products to the user.
- **Unusual activity event:** An unusual real world event attracts users. (e.g., abnormal activity pattern of the elderly, attending Christmas party, and

going out at unusual time) We can provide an effective advertisement related to such unusual events. Many studies attempt to detect outlier events in a time-series [20,13,10].

- **Health indicator:** In addition to activity information, we can measure health indicators with various wearable sensors such as thermometers and pulse monitors. When a user receives health information, an advertisement related to health care products may attract the user.

In the following, we show several activity-based advertising on our sensor-based pervasive systems.

4 Object-Blog

We assume an indoor environment where wireless sensor nodes equipped with such sensors as accelerometers and thermometers are attached to objects. In the Object-blog system [28], the anthropomorphic objects automatically post weblog entries to a weblog about events they experience and their utilization. Fig. 1 shows a weblog entry of temperature changes. The entry includes a movie that shows the temperature changes in a room during one day created by using thermometers attached to objects. (The entry was posted by a virtual object.)

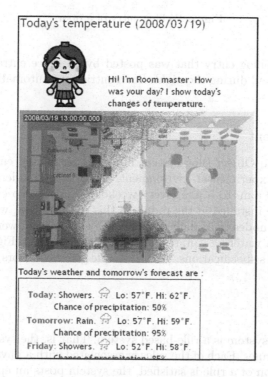

Fig. 1. Weblog entry of temperature changes during one day

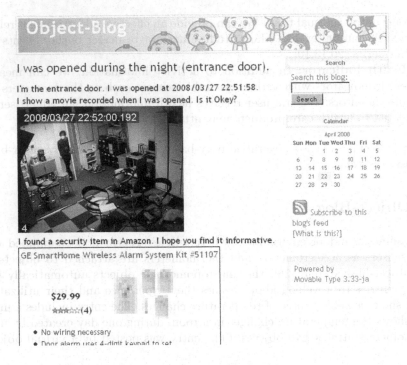

Fig. 2. Weblog entry of a door

Fig. 2 shows a weblog entry that was posted by an office entrance door when the door was opened during the night. Such entries are automatically generated in the weblog.

4.1   Environment

We implement the Object-Blog system in our experimental environment. Fig. 3 (a) shows the experimental environment. Ten workers undertook their ordinary work from 9 a.m. to 5 p.m. every weekday. Eight video cameras and two microphones were installed in the room. In this environment, we taped our implemented sensor nodes to 50 daily objects such as doors, drawers, cups, a coffee maker, a kettle, a watering can, slippers, and a toothbrush. Fig. 3 (b) shows a sensor node and its specifications. We use simple generic sensors to detect object usage.

4.2   Design

The Object-Blog system is a rule-based system. That is, the system has a set of condition-action rules. Each of the rules is associated with a physical object and, when the condition of a rule is satisfied, the system posts an episode about the object associated with the rule. The conditions of the rules are categorized into:

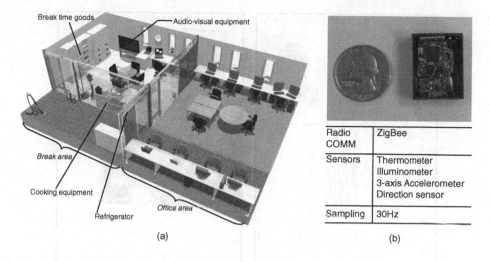

Radio COMM: ZigBee | Sensors: Thermometer, Illuminometer, 3-axis Accelerometer, Direction sensor | Sampling: 30Hz

Radio COMM	ZigBee
Sensors	Thermometer Illuminometer 3-axis Accelerometer Direction sensor
Sampling	30Hz

(a)   (b)

Fig. 3. (a) Experimental environment, and (b) sensor node

periodic (e.g., every Wednesday at 24 o'clock), namely when an event occurs, and when a user makes a request that triggers episode generation. Here, users can request episode generation by posting an entry to the Object-Blog. (For example, a user requests an episode that includes a digest of the clothes the user has worn over the past x-days.) That is, if a user posts an entry that requests episode generation, the requested episode is posted as an entry comment. With the present system, a user has to make a command for episode generation.

4.3 Implementation

We implemented the Object-Blog system that follows the design guideline. We used MovableType as the weblog platform and used the C# language to implement, for example, sensor data processing and automatic episode making. We explain three important components of our implementation.

Event Detection. We implemented four event detection plug-ins in the present system. *Conversation detector* detects conversations from microphone signals. *Physical phenomena detector* detects about ten kinds of physical phenomena such as the movement, dropping, and rotation of objects by using acceleration signals. *Object event detector* detects events that are specific to about ten kinds of objects such as doors, cabinets, and drawers. It detects, for example, the opening and shutting of the objects by using signals from accelerometers and direction sensors. *ADL detector* detects about twenty kinds of activities of daily living (ADLs) such as make tea, cook pasta, cook rice, make juice, brush teeth, listen to music, and practice aromatherapy by machine learning. We implemented the detector based on [44].

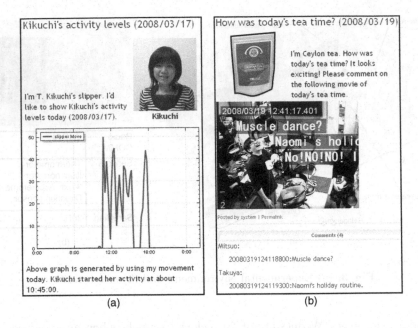

Fig. 4. Weblog entries of (a) activity levels during one day and (b) a comment-able movie

Episode Making. To provide content effectively, we implemented such plug-ins for episode making as the movie plug-in, graph plug-in, and superimposed picture plug-in. The movie in Fig. 1 that represents the changes in temperature is created by using the movie plug-in and superimposed picture plug-in. Fig. 4 (a) shows an entry that includes a graph of a user's activity levels during one day. It is necessary to provide graphs and charts for the easy understanding of a holistic image and for discovering new evidence. We also implemented the comment-able movie plug-in that enables us to annotate entries easily and enjoyably. Fig. 4 (b) shows an entry that encourages us to annotate a conversation during tea time. The entry includes a movie with voice that was recorded during tea time. If a user leaves a comment consisting of time and text on the entry, the text is shown on the movie at the time in the Times Square format, which scrolls the text on one horizontal line across the screen. This function is inspired by the system used by the video-sharing site Nico-nico Video (www.nicovideo.jp). Using the function, we can annotate episodes by sharing and exchanging opinions and experiences.

Advertising on Weblog Entries. We can provide effective advertisements according to weblog entries that relate to a user's daily activities. A weblog entry in Fig. 2 includes an advertisement related to the weblog entry (i.e., security items). Fig. 5 shows another example. In this entry, a CD found that it

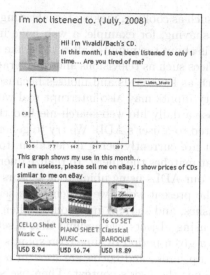

Fig. 5. Weblog entry of a CD

was rarely played, and it provides an advertisement related to resellers. As the above example, we can provide advertisements according to a user's real world activities.

5   Context-Aware Web Page Retrieval

Web pages were originally browsed only on PCs. As a result of device downsizing and advances in wireless communication technology, we are now able to browse web pages anywhere on small devices such as cellphones and PDAs. Nowadays, various home electronic appliances can also be connected to the Internet. Many of the latest televisions are equipped with a LAN port and some of them have a web browser. Google TV[1] and Apple TV[2] that seamlessly connect TVs with the Internet have also become available. Furthermore, an always-on gadget with a web browser named Chumby[3], which is installed in living environments and constantly displays web content, is now being marketed. There will soon come a time when many home electronic appliances connected to the Internet will be embedded in our living environments.

The course of events has convinced us that next-generation web browsers will also be embedded in our living environments. This will enable us to enjoy web pages anywhere and at any time from the browsers around us and obtain useful advice. In particular, we consider that web browsers embedded in electronic appliances will be able to provide useful information related to user's activities

---

[1] Google, Google TV, http://www.google.com/tv/.
[2] Apple Inc., Apple TV, http://www.apple.com/appletv/.
[3] Chumby Industries, Chumby, http://www.chumby.com/.

of daily living (ADLs) such as cooking, making tea, cleaning rooms, and brushing teeth. When a user is shaving, for example, a web page including such tips as 'the best time to shave is about ten minutes after you wake up' may prove useful. However, many appliances such as televisions and refrigerators are not equipped with rich interfaces such as keyboards and mouses because of space limitations and design issues. Query inputs may also interrupt real world activities.

In [29,30], we propose a daily life web search method that automatically retrieves a web page related to a user's ADL. We try to generate queries automatically about ADLs that are currently being undertaken, to search the web, and to show a web page that matches the query. To realize query free searches in our daily lives, we monitor our ADLs using ubiquitous sensors installed in our daily life environments. In the present work, we attach these sensors to daily objects such as cups, toothbrushes, and shavers and monitor their use during daily life. To search the web by using objects, we assume that a set of names of objects employed by a user in a given period of time corresponds to the user's context. For example, when a cup, milk, and cocoa are used in a given period of time, "cup milk cocoa" becomes the user's context. Then, we search for a web page that matches that context. That is, we form a query/queries from the context with which to search the web. The strategy behind the method reflects our idea that 'a web page including the names of objects that are used in an ADL may relate to the ADL.'

The method enables us to provide information (web pages) related to a user's ADLs in a timely and effective manner by displaying it on web browsers embedded in her daily living environment. We consider that providing ADL-related web pages may have the following benefits.

1. As with the above example of shaving, a user can access information that enriches her daily life and improves her activities.
2. The method can provide a user with amusing and useful information about ADLs that can trigger communication with other users.
3. As shown by the popularity of trivia related books and TV shows, many users gain satisfaction from obtaining background knowledge about items and phenomena in their lives. Our method can satisfy this desire for knowledge about daily living.
4. The method can provide us with information (including advertisement) about a ADL product. Obtaining the information may convince us to buy the product.

5.1   Our Implemented Web Browser

We introduce our implemented web browsers for televisions. We call a web browser implemented on a television a *TV browser*. We install the browser on a Windows PC and connect its video-out to a television. We install a TV tuner on the PC to make it possible to view TV programs on the PC. We select Nintendo Wii Remote as the input interface of the browser. One of the remote features of Wii is a function for pointing at a TV screen. Wii has already achieved high sales, and so its

Fig. 6. TV browser with a ticker

pointing function is being accepted by home users. We connect a Wii Remote and the PC via a Bluetooth adapter so that we can operate the mouse cursor of the PC by employing Wii Remote's pointing function. The Wii Remote also has several buttons and a directional pad. The directional pad is used for manual page scrolling. The TV browser presents a web page of tips and trivia related to the user's ADLs, e.g., making tea, watering plants, and vacuuming.

Our assumed web browsers are embedded in living environments. Information terminals that are embedded in real environments and presents non-critical information are called peripheral displays [32], notification systems [33], ambient displays [12,32], and ambient information systems [37][4]. Many of these systems are designed so that they do not disturb the user too much. Our web browsers, which are installed in living environments, should also be designed to be unobtrusive. Thus, we simply use sound effect notifications when ADL-related pages are presented. This permits the participants to ignore a web page that is presented when they are busy.

Some peripheral displays provide overviews and detailed views of presented information. For example, Scope [46], which runs on a PC, is a notification system drawn from multiple sources of information, including e-mail, instant messaging, and appointments. Icons that represent notifications are placed on its circular radar-like screen. Because each icon shows the properties of its information, e.g., the importance of an e-mail, the icon permits a user to obtain an overview of the information easily. A user can also easily access the information, e.g., the main body of an e-mail, by simply double-clicking the icon. We consider that a web browser on an electronic appliance should also enable a user to obtain an overview of a web page easily simply by focusing on the browser. In the pilot usability study, we found that many participants wanted to obtain the contents of web pages with a minimum of effort. We show two example cases. First, a participant said that she sometimes had to stare hard at the display to understand the context of the page. It was difficult for participants to understand the context of the page easily at a glance who were not in front

---

[4] Note that the strict definitions of the display (system) differ slightly for each term and in each paper. For more detail, see [37].

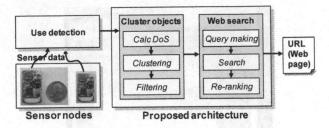

Fig. 7. Architecture of web search for daily living

of the display when the display presented the page. Also, when an ADL-related page is presented, a user usually cannot access the information easily because she is performing an ADL. Second, some participants said that picking up the remote is troublesome. When the main content of a page was not located in the top portion of the page, for example, the participants sometimes had to scroll the page manually to find it. We consider that users should be able to access an overview of the presented page at a glance. Then, when a user is interested in the page, she can examine it in detail by picking up the remote. Therefore, in addition to the sound effect notification, we use a ticker, that is, the title and the introduction of the page is shown in the Times Square format, which scrolls the text across one horizontal line in the center of the display over the web page display area as shown in Fig. 6.

5.2   Daily Life Web Search Method

Here we try to retrieve ADL-related web pages automatically without the need for any metadata or labeled training data. The daily life web search method proposed in [29] exploits the use of daily objects involved in an ADL detected by object attached sensors. By employing the use and names of objects, we generate a query designed to retrieve an ADL-related web page. Fig. 7 shows the architecture of the web search method for daily living. The method detects the use of objects in the past t minutes for each object every t minutes. It searches the web by employing the use detected in t minutes. The method consists of two main procedures: '*Cluster objects*' and '*Web search*.' In '*Cluster objects*,' we generate clusters consisting of objects that are simultaneously used in the same ADL in t minutes. Sometimes users employ (move) objects that are not directly related to the ADL they are performing. When making tea, for example, a user may move not only tea but also cocoa or salt, which are unrelated to making tea, but that happen to be in the same cabinet. Furthermore, two or more people may simultaneously perform different ADLs in a given environment. For example, one person may be shaving with a razor and another person may be using a toothbrush and toothpaste. To cope with these issues, we cluster objects by type of ADL. For example, as regards brushing, we group such objects as a toothbrush and toothpaste as a cluster.

In '*Web search*,' we search the web by using the names of objects in the same cluster to obtain a web page for each cluster. In this implementation, we set t at three. In a preliminary experiment, the average ADL time period during which participants used multiple objects was about two minutes, therefore we assume that object use in a three-minute period yields sufficient accuracy. After explaining how we detect the use of objects, we explain *Cluster objects* and *Web search* in detail.

Use Detection. We assume that sensor nodes are attached to daily objects to monitor their use. We use our developed sensor node, which is equipped with a three-axis accelerometer that obtains its acceleration every 30 msec and sends wirelessly it to a data server. Fig. 3 (a) shows our developed sensor node. We assume that a time period during which acceleration data change greatly coincides with the period during which an object is used. Because we use a three-axis (x, y, and z) accelerometer, the combined time periods of change extracted from the data of each axis sensor are defined as the time periods during which an object is used. Some signal processing studies use the Gaussian Mixture Model to learn the Fourier components of the change and the noise (no-change) time periods [43]. We adopt a similar approach to detect changes.

Cluster Objects. '*Cluster objects*' consists of three procedures in Fig. 7. We call the measure for determining whether two objects are used in the same ADL the 'Degree of being used in the *Same* ADL (*DoS*).' We assume that the 'probability' with which two objects are used in the same ADL increases as the *DoS* of the objects increases. We compute the *DoS* for all the objects that are used in the time period (*Calc DoS*) and cluster the objects according to the *DoS* (*Clustering*). We then filter the clusters including only objects that are used for a short time in the time period (*Filtering*).

We explain how to calculate the *DoS*. This procedure computes *DoS* between objects X and Y in time period t (three minutes) based on the following three measures.

- $Temp(X, Y, t)$: This measure represents the degree to which X and Y are used simultaneously in t. It uses the activities of X and Y in t. This is based on the assumption that simultaneously used objects are employed in the same ADL with high probability.
- $Hist(X, Y)$: This measure represents the degree to which X and Y are used simultaneously in a past dataset. We prepare a past dataset of a certain period in advance. This measure reflects the view that the frequency with which objects were simultaneously used in the past dataset increases as the probability with which the objects are used in the same ADL increases.
- $Sem(X, Y)$: This measure represents the semantic relevance between X and Y. We use the co-occurrence of X and Y in the web as the semantic relevance. This is based on the view that objects that are ordinarily used together in the real world co-occur in the web, which reflects the real world. There are

two major approaches for computing the co-occurrence of two words. The first employs page counts (hit counts) obtained from a search engine by using queries involving the words. The second employs the co-occurrence of the two words within a text window in web pages. We employ the first approach because it requires no parameters and we can easily compute the co-occurrence by using web search APIs that provide hit count information.

We do not provide detailed formula of three measures because this section mainly focuses on web page retrieval. By using the measures, the DoS is represented as follows:

$$DoS(X, Y, t) = Temp(X, Y, t) \cdot Hist(X, Y) \cdot Sem(X, Y).$$

Here we briefly describe *Clustering* and *Filtering*. In *Clustering*, we use Ward's method [9] to cluster objects. We use the DoS as inverse distance between objects. After that, we eliminate outliers from each cluster as a postprocessing. *Filtering* eliminates clusters that have no object whose total seconds of activity in t exceed a threshold ε_{ac}. We set $\varepsilon_{ac} = 5$.

Web Search. The *Cluster objects* procedure produces clusters that consist of a set of objects. The cluster corresponds to user's context. We construct subqueries from each cluster and obtain web pages corresponding to the query. As shown in Fig. 7, this procedure consists of three procedures. First, we build multiple subqueries from a cluster (*Query making*) and obtain multiple search results for the subqueries (*Search*). Then, we re-rank the search results by computing the similarities between each of the web pages and the cluster (*Re-ranking*) and output the URL of the top-1 page in the re-ranked list.

Query making uses a vector representation of a cluster. In the representation, a cluster is represented in a vector that reflects the 'importance' of an object in the cluster. The importance of an object is simply defined as the largest DoS between the object and another in the cluster. The importance shows the contribution of the object used in its corresponding ADL. That is, it reflects the semantic relationship between objects, whether an object is simultaneously used with another object, and an object use in a past dataset. *Query making* first produces a list (vector) of objects with their importance such as <juicer,3.0>, <cup,3.0>, <milk,2.0>, <cup,1.0>, and <sugar,0.5>. If a list includes multiple objects with the same name, we remove the duplicate objects except for the object with the largest importance. <cup,1.0> is removed in the above example and a four-dimensional vector is obtained. We call the vector a **context vector**. The *Web search* method finds a web page that is related to an ADL by using a context vector. *Query making* applies the following three techniques to a context vector in order.

[**1. Vector expansion**]: Assume that a user made tea using green tea and a cup in a given time period. She then drank the tea using only the cup after that time period. Retrieving web pages that relate to drinking green tea is difficult by using a query constructed solely from the cup. However, we can easily retrieve web pages related to green tea by building a query from the 'green tea' that was used with the cup. We thus expand a current context vector V_i by using vectors that were constructed in the past. Let us define a similarity $Sim_h(V_i, V_j)$ between vectors V_i and V_j as follows.

$$Sim_h(V_i, V_j) = \lambda^{d(V_i, V_j)} \cos(V_i, V_j),$$

where $\cos(V_i, V_j)$ is a cosine similarity between V_i and V_j; $d(V_i, V_j)$ is the temporal distance (minutes) between time periods of V_i and V_j; λ ($0 < \lambda \le 1$) is a forgetting factor [31] that controls the effect of past values. Assume that the number of dimensions (the number of objects) of a context vector V_i is equal to or less than a threshold ε_{hq}. Then the procedure *Vector expansion* finds a past context vector V_j that satisfies the following conditions; $Sim_h(V_i, V_j)$ has the maximum value among past vector similarity values that are greater than a threshold ε_{hw}; not all the dimensions of V_j are identical to those of V_i. Otherwise the procedure does nothing. After finding V_j, the procedure expands V_i by multiplying V_j by $\lambda^{d(V_i, V_j)}$, and then adding it to V_i. In our implementation, we set $\varepsilon_{hq} = 2$, $\varepsilon_{hw} = 0.7$, and $\lambda = 0.99$.

[**2. Making subqueries**]: The procedure *Making subqueries* makes multiple subqueries by extracting some objects (terms) from a vector. By extracting objects, we can construct some subqueries that are not too strict and do not include noises, i.e., the names of objects wrongly included in the cluster constructed in the *Cluster objects* procedure. Before the construction, we compute the inverse document frequency (idf) of an object name for each dimension of the context vector. We use the function $log(D/f(X))$, where $f(X)$ is the frequency of the object X's name in the document collection and D is the number of documents, to compute the idf. Because we use Yahoo! web search[5] to compute $f(X)$, we set D as 20 billion. Then, we multiply the importance of each dimension by its idf. This multiplication can decrease importance of objects whose name appears frequently in web documents such as cup.

We simply construct subqueries from all combinations of objects in a context vector. Assume that a context vector has i objects and the desired query length is l. We can make $_iC_l$ subqueries. Also, to limit the number of subqueries, we discard subqueries that do not include an object with a top-s importance. When $l = 2$ and $s = 2$, the example context vector (juicer, cup, milk, and sugar) shown at the beginning of this section produces subqueries "juicer cup," "juicer milk," "juicer sugar," "cup milk," and "cup sugar." That is, this algorithm outputs combinations of objects while giving priority to more important objects. We set $s = 2$ in our implementation.

[**3. Query expansion**]: A query that includes only the names of objects may be ambiguous. That is, it is difficult to obtain our desired pages (daily life information) from such a query as "cup green-tea." In [14], it is reported that combining a topic related term and a genre related term makes a good query. If we want to buy a camera, for example, we combine the topic term "camera" and the genre term "buying" or "choosing," i.e., we make the query "camera buying." We also expand the query by using genre terms that may yield web

---

[5] Yahoo! developer network, http://developer.yahoo.com.

pages about daily lives. This algorithm randomly selects a genre term from the four terms ("advice," "how-to," "tips," and "trivia").

We can obtain multiple search results (rankings) by sending multiple sub-queries constructed in *Query making* to a search engine. The *Re-ranking* is performed by using a similarity measure between a context vector and a web page in the search results to combine the multiple rankings into one ranking. The top-1 URL (page) in the re-ranked list is produced. We have prepared five re-ranking algorithms based on commonly used (re-)ranking algorithms in context search and/or meta search, and then evaluate the algorithms by using real data set. For more detail, refer to [29].

5.3   Advertising on Context-Aware Web Page Retrieval

By using the above method, we can find a web page that relates to a user's activity. As mentioned in section 2, many studies try to find an advertisement that relates to a web page. That is, by using such studies (techniques), we can provide an advertisement that relates to a web page presented on the above system. Also, as we mentioned above, our method can provide the user with information about a ADL product. Obtaining the information may convince the user to buy the product.

6   Conclusion

This article describes a new context-aware advertising that employs daily activity information captured by ubiquitous sensors. By employing ubiquitous sensors, we can provide an effective advertisement that relates to the user's current activity and detailed situation. We presented two example applications that employ context information captured by ubiquitous sensors, and described how we provide advertisements on the example applications. As a part of our future work, we plan to embed an advertising function in our TV Browser.

References

1. Adomavicius, G., Sankaranarayanan, R., Sen, S., Tuzhilin, A.: Incorporating contextual information in recommender systems using a multidimensional approach. ACM Transactions on Information Systems (TOIS) 23(1), 103–145 (2005)
2. Baeza-Yates, R., Ribeiro-Neto, B.: Modern information retrieval. Addison-Wesley (1999)
3. Bahl, P., Padmanabhan, V.: RADAR: An in-building RF-based user location and tracking system. In: IEEE INFOCOM 2000, vol. 2, pp. 775–784 (2000)
4. Bao, L., Intille, S.S.: Activity recognition from user-annotated acceleration data. In: Ferscha, A., Mattern, F. (eds.) PERVASIVE 2004. LNCS, vol. 3001, pp. 1–17. Springer, Heidelberg (2004)
5. Blum, M., Pentland, A., Tröster, G.: Insense: Interest-based life logging. IEEE Multimedia 13(4), 40–48 (2006)

6. Clarkson, B., Mase, K., Pentland, A.: Recognizing user context via wearable sensors. In: Int'l Symp. on Wearable Computers, pp. 69–75 (2000)
7. Dhar, S., Varshney, U.: Challenges and business models for mobile location-based services and advertising. Communications of the ACM 54(5), 121–128 (2011)
8. Gemmell, J., Bell, G., Lueder, R., Drucker, S., Wong, C.: MyLifeBits: Fulfilling the Memex vision. In: ACM Multimedia 2002, pp. 235–238 (2002)
9. Hair, J., Anderson, R., Tatham, R., Black, W.: Multivariate data analysis
10. Hodge, V., Austin, J.: A survey of outlier detection methodologies. Artificial Intelligence Review 22(2), 85–126 (2004)
11. Huang, Z., Zeng, D., Chen, H.: A comparison of collaborative-filtering recommendation algorithms for e-commerce. IEEE Intelligent Systems 22(5), 68–78 (2007)
12. Ishii, H., Ren, S., Frei, P.: Pinwheels: Visualizing information flow in an architectural space. In: CHI 2001 Extended Abstracts, pp. 111–112 (2001)
13. Jain, A., Chang, E., Wang, Y.: Adaptive stream resource management using Kalman filters. In: SIGMOD Conference 2004, pp. 11–22 (2004)
14. Kraft, R., Stata, R.: Finding buying guides with a web carnivore. In: 1st Latin American Web Congress (LA-WEB), pp. 84–92 (2003)
15. Kurkovsky, S., Harihar, K.: Using ubiquitous computing in interactive mobile marketing. Personal and Ubiquitous Computing 10(4), 227–240 (2006)
16. LaMarca, A., Chawathe, Y., Consolvo, S., Hightower, J., Smith, I., Scott, J., Sohn, T., Howard, J., Hughes, J., Potter, F., Tabert, J., Powledge, P.S., Borriello, G., Schilit, B.N.: Place lab: Device positioning using radio beacons in the wild. In: Gellersen, H.-W., Want, R., Schmidt, A. (eds.) PERVASIVE 2005. LNCS, vol. 3468, pp. 116–133. Springer, Heidelberg (2005)
17. Langheinrich, M., Nakamura, A., Abe, N., Kamba, T., Koseki, Y.: Unintrusive customization techniques for web advertising. Computer Networks 31(11), 1259–1272 (1999)
18. Lester, J., Choudhury, T., Borriello, G.: A practical approach to recognizing physical activities. In: Fishkin, K.P., Schiele, B., Nixon, P., Quigley, A. (eds.) PERVASIVE 2006. LNCS, vol. 3968, pp. 1–16. Springer, Heidelberg (2006)
19. Linden, G., Smith, B., York, J.: Amazon.com recommendations: Item-to-item collaborative filtering. IEEE Internet Computing 7(1), 76–80 (2003)
20. Liu, H., Shah, S., Jiang, W.: On-line outlier detection and data cleaning. Computers & Chemical Engineering 28(9), 1635–1647 (2004)
21. Lukowicz, P., Junker, H., Stäger, M., von Büren, T., Tröster, G.: WearNET: A distributed multi-sensor system for context aware wearables. In: Borriello, G., Holmquist, L.E. (eds.) UbiComp 2002. LNCS, vol. 2498, pp. 361–370. Springer, Heidelberg (2002)
22. Lukowicz, P., Ward, J.A., Junker, H., Stäger, M., Tröster, G., Atrash, A., Starner, T.: Recognizing workshop activity using body worn microphones and accelerometers. In: Ferscha, A., Mattern, F. (eds.) PERVASIVE 2004. LNCS, vol. 3001, pp. 18–32. Springer, Heidelberg (2004)
23. Maekawa, T., Kishino, Y., Sakurai, Y., Suyama, T.: Recognizing the use of portable electrical devices with hand-worn magnetic sensors. In: Lyons, K., Hightower, J., Huang, E.M. (eds.) PERVASIVE 2011. LNCS, vol. 6696, pp. 276–293. Springer, Heidelberg (2011)
24. Maekawa, T., Kishino, Y., Yanagisawa, Y., Sakurai, Y.: Mimic sensors: Battery-shaped sensor node for detecting electrical events of handheld devices. In: Kay, J., Lukowicz, P., Tokuda, H., Olivier, P., Krüger, A. (eds.) PERVASIVE 2012. LNCS, vol. 7319, pp. 20–38. Springer, Heidelberg (2012)

25. Maekawa, T., Kishino, Y., Yanagisawa, Y., Sakurai, Y.: Recognizing handheld electrical device usage with hand-worn coil of wire. In: Kay, J., Lukowicz, P., Tokuda, H., Olivier, P., Krüger, A. (eds.) PERVASIVE 2012. LNCS, vol. 7319, pp. 234–252. Springer, Heidelberg (2012)
26. Maekawa, T., Watanabe, S.: Unsupervised activity recognition with user's physical characteristics data. In: Int'l Symp. on Wearable Computers, pp. 89–96 (2011)
27. Maekawa, T., Yanagisawa, Y., Kishino, Y., Ishiguro, K., Kamei, K., Sakurai, Y., Okadome, T.: Object-based activity recognition with heterogeneous sensors on wrist. In: Floréen, P., Krüger, A., Spasojevic, M. (eds.) PERVASIVE 2010. LNCS, vol. 6030, pp. 246–264. Springer, Heidelberg (2010)
28. Maekawa, T., Yanagisawa, Y., Kishino, Y., Kamei, K., Sakurai, Y., Okadome, T.: Object-blog system for environment-generated content. IEEE Pervasive Computing 7(4), 20–27 (2008)
29. Maekawa, T., Yanagisawa, Y., Sakurai, Y., Kishino, Y., Kamei, K., Okadome, T.: Web searching for daily living. In: SIGIR 2009, pp. 27–34 (2009)
30. Maekawa, T., Yanagisawa, Y., Sakurai, Y., Kishino, Y., Kamei, K., Okadome, T.: Context-aware web search in ubiquitous sensor environment. ACM Transactions on Internet Technology (ACM TOIT) 11(3), 12:1–12:23 (2012)
31. Markovitch, S., Scott, P.: The role of forgetting in learning. In: ICML 1988, pp. 459–465 (1988)
32. Matthews, T., Rattenbury, T., Carter, S., Dey, A., Mankoff, J.: A peripheral display toolkit. In: UIST 2004, pp. 247–256 (2004)
33. McCrickard, D., Chewar, C.: Attentive user interfaces: Attuning notification design to user goals and attention costs. Communications of the ACM 46, 67–72 (2003)
34. Mihailidis, A., Carmichael, B., Boger, J.: The use of computer vision in an intelligent environment to support aging-in-place, safety, and independence in the home. IEEE Trans. on Info. Tech. BioMedicine 8(3), 238–247 (2004)
35. Mynatt, E., Rowan, J., Craighill, S., Jacobs, A.: Digital family portraits: Supporting peace of mind for extended family members. In: CHI 2001, pp. 333–340 (2001)
36. Philipose, M., Fishkin, K., Perkowitz, M.: Inferring activities from interactions with objects. IEEE Pervasive Computing 3(4), 50–57 (2004)
37. Pousman, Z., Stasko, J.: A taxonomy of ambient information systems: Four patterns of design. In: Int'l Working Conference on Advanced Visual Interfaces (AVI 2006), pp. 67–74 (2006)
38. Prasithsangaree, P., Krishnamurthy, P., Chrysanthis, P.: On indoor position location with wireless LANs. In: IEEE International Symposium on Personal, Indoor and Mobile Radio Communications, vol. 2, pp. 720–724 (2002)
39. Ravi, N., Dandekar, N., Mysore, P., Littman, M.: Activity recognition from accelerometer data. In: IAAI 2005, vol. 20, pp. 1541–1546 (2005)
40. Ribeiro-Neto, B., Cristo, M., Golgher, P., de Moura, E.: Impedance coupling in content-targeted advertising. In: SIGIR 2005, pp. 496–503 (2005)
41. Salton, G.: Automatic Text Processing. Addison-Wesley (1989)
42. Shi, Y., Huang, Y., Minnen, D., Bobick, A., Essa, I.: Propagation networks for recognition of partially ordered sequential action. In: CVPR 2004, vol. 2, pp. 862–869 (2004)
43. Sohn, J., Kim, N., Sung, W.: A statistical model-based voice activity detection. IEEE Signal Processing Letters 6(1), 1–3 (2002)
44. Tapia, E.M., Intille, S.S., Larson, K.: Activity recognition in the home using simple and ubiquitous sensors. In: Ferscha, A., Mattern, F. (eds.) PERVASIVE 2004. LNCS, vol. 3001, pp. 158–175. Springer, Heidelberg (2004)

45. Munguia Tapia, E., Intille, S.S., Larson, K.: Portable wireless sensors for object usage sensing in the home: Challenges and practicalities. In: Schiele, B., Dey, A.K., Gellersen, H., de Ruyter, B., Tscheligi, M., Wichert, R., Aarts, E., Buchmann, A.P. (eds.) AmI 2007. LNCS, vol. 4794, pp. 19–37. Springer, Heidelberg (2007)
46. van Dantzich, M., Robbins, D., Horvitz, E., Czerwinski, M.: Scope: Providing awareness of multiple notifications at a glance. In: Int'l Conf. on Advanced Visual Interfaces (AVI 2002), pp. 157–166 (2002)
47. van Kasteren, T., Noulas, A., Englebienne, G., Kröse, B.: Accurate activity recognition in a home setting. In: Ubicomp 2008, pp. 1–9 (2008)
48. Wu, J., Osuntogun, A., Choudhury, T., Philipose, M., Rehg, J.: A scalable approach to activity recognition based on object use. In: ICCV 2007, pp. 1–8 (2007)
49. Yih, W., Goodman, J., Carvalho, V.: Finding advertising keywords on web pages. In: WWW 2006, pp. 213–222 (2006)
50. Yuan, S., Tsao, Y.: A recommendation mechanism for contextualized mobile advertising. Expert Systems with Applications 24(4), 399–414 (2003)

Towards Early Detections of the Bad Debt Customers among the Mail Order Industry

Masakazu Takahashi[1] and Kazuhiko Tsuda[2]

[1] Graduate School of Innovation &Technology Management, Yamaguchi University
2-16-1, Tokiwadai, Ube, Yamaguchi 755-8611, Japan
[2] Graduate School of Business Sciences, University of Tsukuba
3-28-1, Otsuka, Bunkyo, Tokyo 120-0011, Japan
masakazu@yamaguchi-u.ac.jp, tsuda@gssm.otsuka.tsukuba.ac.jp

Abstract. This paper presents investigating the customer characteristics of mail order industry, especially the bad debt customers. These kinds of investigations have not made intensively, such as private default risks so far and conventional method for predicting such risks depend on the employee's working experiences. For these backgrounds, we observed the bad debt list gathered from a mail order company and analyzed combined with the sales data. From the results of the research, we characterized the potential bad debt customers with the machine learning method. Intensive research has revealed that the characteristics of customers who might fall into the bad debt list. This result will make use for the revenue expansion with the improvement of the bad debts collection.

Keywords: Mail Order, Customer Analysis, Bad Debts, Random Forest, Machine Learning, Service Science and Management Engineering.

1 Introduction

Mail order industry is one of the promising methods of sales expansion even such as deflation condition overlap with long-term slump in Japanese retail business that has characteristics of delivering the items to the customers' hand. A survey from the Japan Direct Marketing Association says the percentage of credit losses of a mail order company is estimated about 0.5% of net sales. It is important for the mail order industry to predict risk exposures in customers' credit control. It is because too large risk exposure leads to high default risk and too small risk exposure misses business opportunities. Therefore, in the mail order industry, the credit-scoring strategy is one of the serious problems for the further sales expansion. In this paper, we will carry out data analysis for the purpose of the customer behavior through the bad debt list from a mail order company.

The rest of the paper is organized as follows: Section 2 discusses the backgrounds of the research and related works; Section 3 briefly summarize the gathered data on the target mail order company; Section 4 describes analytics of the data and presents analytical results; and Section 5 gives some concluding remarks and future works.

T. Matsuo & R. Colomo-Palacios (Eds.): *Electronic Business and Marketing*, SCI 484, pp. 167–176.
DOI: 10.1007/978-3-642-37932-1_12   © Springer-Verlag Berlin Heidelberg 2013

2   Backgrounds and Related Works

While the domestic retail business is in the long term slump, the mail order industry is continuing the sales expansion. One of the conclusive factors of the expansion is the reversionary payment method. Meanwhile, it increases in number such as fraud transactions from vicious customers in recent year. Therefore, it is important to realize safe dealings in the mail order industry, that provides redistribution to the customers and reduce sunken security costs. Moreover, to increase the transactional number of the mail order that is achieving economic growth in Japan, let the domestic economic demand stimulate.

According to the mail order industry sales survey, it was estimated amount to 4,310 billion yen in total of the 2009 fiscal year which increased in 170 billion yen of 4.1% raise compared with previous year and became the maximum record since the research starts [1]. One of the factors for sales expansion among the mail order industry is the diversification of payment methods. Not only the reversionary system that is settles accounts after the order item arrives with which improving customers' conveniences also appeal the transactional safety from the view point of the customers which expect the large increase both of the sales and the profits for the mail order company.

On the other hand, there are no credit criteria at the time of start trading. Therefore, there are many transaction of scam or fraud by vicious customers making use of the reversionary systems. Therefore, we will investigate the overview of the mail order industry at first.

Table1. shows the component of payment methods in the mail order industry [1]. Especially, the percentage of the customers who used the convenience store or the postal transfer as the reversionary systems reached 39.7% in 2009. As far as these reversionary payment systems continue, the fraud transactions will increase in number.

Table 1. The Primary Payment Methods

Payment Methods	2006	2007	2008	2009
Cash On Delivery	26.8%	25.8%	27.6%	28.3%
Convenience Store	24.6%	24.5%	24.2%	23.8%
Credit Card	19.2%	19.7%	21.9%	23.2%
Postal Transfer	21.9%	19.3%	18.5%	15.9%
Bank Transfer	6.7%	9.7%	7.5%	8.2%
Others	0.7%	1.0%	0.3%	0.5%
No Answer	0.2%	0.0%	0.1%	0.1%
Total	100.0%	100.0%	100.0%	100.0%

Next, we describe the issues of the mail order industry from interviewing.

1) Credit Control
Even though about 40 percent of the transactions were selected the reversionary payment method, there are no credit criteria with the new customers just like

information center from the credit card industry in Japan. This issue led continued the fraud transactions.

2) Delivery

Even if the ordered packet has some address difference, the logistics companies correct them and deliver to the recipients. Because, they are aiming for delivering the packets to the recipients. This is the one of the merits for the Japanese logistics companies, whereas it is sometimes used for the confidence games. These two issues originate the hotbed of the fraud transactions.

Concerning to the payment collection, since each payment is petty, they have to take into consideration for trade-off between the collection cost and the invoice. Therefore, they are taking a financial policy for counting up the allowances for the future bad debts as the advance risk hedge for the bad debts. As long as they are taking the reversionary system as one of the diversification of paying methods to make their market expand, there need to understand of the characteristics of the bad debts customers.

Next, we consider how the fraud transactions make from the workflow of the shipment. This is also summarized from the interviewing. Orders that received by 8:00 are processed and forward to the warehouse for preparing for shipping in the morning and then prepared for shipping in the afternoon. Generally, the operators are inspecting visually the data sorted by such as the postal code.

Form the number of the order that is amount to around ten thousands, it is hard to share the time for the inspection with one packet less than a minute. However, it is difficult to distinguish the address in a short time created on the assumption that it was made to mistake and to discover an inaccurate addresses. So far, the working experience of the staff plays an important role for the detection. Therefore, they need to have some support systems for the fraud transactions to increase the detection.

Next, the scale of the amount of bad debt is explored from the public data of a listed mail order companies in Tokyo Stock Exchange. Table2. shows the rate of the amount of bad debt to the sales of listed mail order companies [2]. This figure gathered from sales and general administrative expenses, a non-operating expenditure, and extraordinary loss were considered as the amount sum of bad debt. Moreover, indirect costs to delinquent debts, such as remind calls, reissue of the invoices.

Table 2. Sales and Bad Debts of the Listed Mail Order Companies in Japan

	Scroll	Sensyukai	Nissen	Belluna
Listed Stock Code	8005	8165	8248	9997
Fiscal Period	2012.3	2011.12	2011.12	2012.3
Turn Over (in mil.JPY) (A)	48,712	79,725	4,513	85,432
Bad Debts (in mil.JPY) (B)	450	842	47	747
(B)/(A) %	0.9%	1.1%	1.0%	0.9%

As for the related works on the mail order industry, it separates into the activity of before order and of after order received. The related works of the before order activities were focused on the elaboration of the order received in consideration of the time lag from customers such as the demand predictions [3]-[6].

On the other hand, at the phase for the after order received, most of the researches have been made for customer analysis with the purchase history. There are lots of customer analyses with the data-mining methods [7], [8]. For example, the order history has been used for such as trend analyses of customers and sales promotion strategies as the retailers' decision support tools. As long as the reversionary system is adopted in mail order industry, it is important to collect the payment from the customers as soon as possible, whereas the related works on the collecting the payment was less studied than that of customer analysis, especially collecting the bad debts. Concerning to the research on the bad debts, most of them have been done about the bankruptcy prediction of the corporation [9]-[11]. These are used for the preliminary screening to hedge risks with the machine learning methods [12]-[15].

Among the machine learning methods, the random forest is one of the promising methods for such an insufficient data prediction such as authorship identifications, software production workload estimations, promising customers' classifications, and credit risk evaluations [16]-[22].

From the backgrounds and the related works, we found out that the customer relation management among the mail order industry is mainly used for the sales expansion so far, and one of the characteristics of the mail order system is delivered to the customer hand even both the name and the address is written. Furthermore, most of the customer lists from the mail order company are not fulfilled all the customer attributes.

Therefore, the random forest is one of the promising methods to classify the target customers.

3   Data Summary

To understand current condition of the bad debt of the mail order, we make an intensive research for the bad debt data gathered from the mail order company in Japan. This data was financially processed into the bad debts in March 2011.

Table3. shows the summary for the bad debts. Grand total of the debts amount to $350,571.94 and composed of the 9,357 customers. Average sales price per customer amount to $235.57, Average sales price per item reached $109.49, and Average sales price per transaction scored $148.68 respectively.

Table 3. Debts Data Summary

Fiscal Year	2010
Number of Customers	9,357
Number of Invoices	20,132
Average Sales Price per Customer	$235.57
Average Sales Price per Item	$109.49
Average Sales Price per Transaction	$148.68
Grand total of the bad debt	$350,571.94

Both Table4. and Table5. indicate the proportion and number from the debt customer attributes by sex and age. From the results, around the four thirds of the customers are female with around 40's. This result indicates the image of the primary customer. Meanwhile, there should be suspected the age replies that are both less than 10 years old and over 80 years old. It is hard to imagine that those who the age below 10 or over 80 could purchase items through the mail order without supporting the other person.

Table 4. Customer Attributes from the Bad Debts by Sex

	Number	%
Male	2,443	26.11%
Female	6,894	73.68%
N.A.	20	0.21%
Total	9,357	100.00%

Table 5. Customer Attributes from the Bad Debts by Age

	Min.	Max.	Ave.	Answered	N.A.
Male	5	90	48.13	2,084	359
Female	3	97	42.62	5,960	934
N.A.	22	88	48.25	12	8
Total	3	97	44.05	8,056	1,301

Both Table6. and Table7. indicate the number of transactions by category and the amount of debts by category, respectively. Both tables indicate that Lady's items shared majority.

Table 6. Number of the Transaction by the Categorical Bad Debt Classification

Category	Transaction	%
Lady's	10,629	52.80%
Men's	1,198	5.95%
Others	8,305	41.25%
Total	20,132	100.00%

Table 7. Amount of the Bad Debts by Category

Category	Debts	%
Lady's	$1,078,040.20	48.91%
Men's	$94,517.18	4.29%
Others	$1,031,640.08	46.80%
Total	$2,204,197.47	100.00%

From the basic research from the bad debts list, we have figured out the characteristics of the attributes and the item categories. In the system of the mail order industry, the ordered items will deliver to the customer hands at least both the name and the address is written exactly. Therefore, the mail order company needs to predict the potential customer who might fall into the dad debts list.

4   Data Analysis

Since this gathered data is not cover all the transaction, we focus on the percentage of transaction and customer per 10,000 populations by prefecture in order to figure out the local characteristics at first. Concerning to the population, we make use of Census as of October 1st, 2010 [22]. Fig.1 indicates propotion of debt transaction per 10,000 populations by prefecture and Fig.2 also indicates propotion of debt customer per 10,000 populations by prefecture respectively. From both figures, Hokkaido, Osaka, and Fukuoka scored high ratio of bad debt both of the transaction and customer compared with another prefecture. Based on the past studies and the characteristics of the machine learning, we will make use of the random forest to characterize the bad debt customers with this insufficient list.

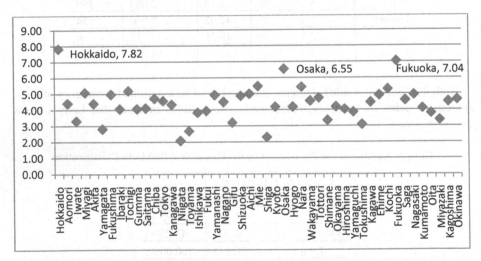

Fig. 1. Percentage of Debt Transaction per 10,000 Population by Prefecture

Random forest has some merits as we mentioned in the related works. We have summarized the bad debt list into 20,132 records by order date and customer ID. We will make characterization five type customer conditions such as 1) claim difficulty, 2) collecting difficulty, 3) death with no inheritance, 4) missing and 5) personal bankruptcy with the following eleven attributes; a) Customer ID, b) Prefecture, c) Sex, d) Age, e) Debt Processing Period, f) Payment Method, g) Number of Payment, h) Number of Remittance, i) Phone Call Status, j) Number of Purchased Items, k)

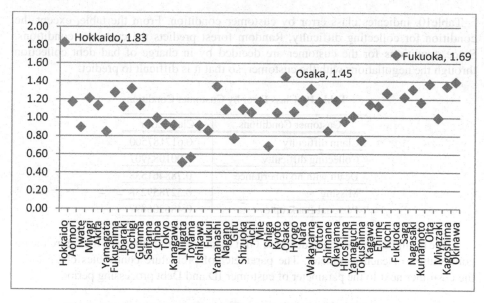

Fig. 2. Percentage of Debt Customer per 10,000 population by Prefecture

Table 8. Data Configration

Data	Records	%
Test data	12,079	60%
Training data	8,053	40%
Total	20,132	100%

Amount of Sales. Among the summarized data, we set sixty percent of data as training data (12,079 records) and lest of forty percent of the data as test data (8,053 records) with random number assignment in Table 8.

As a result, 500 trees from 3 parameters classified customer conditions from 6 parameters with class errors in table9. The error rate for the test with OOB (out-of-bag) data results 4.97%. This score means that the cluster calculated with the classifier scored 4.97% error compared with the cluster contained in the original data. OBB data is used to get a running unbiased estimate of the classification error as trees are added to the forest. It is also used to get estimates of variable importance.

Table 9. Result of Random Forest Trial

Type of random forest	Classification
Number of trees	500
Number of variables tried at each split	3
OOB estimate of error rate	4.97%

Table10. indicates class error by customer condition. From the table, except the condition for collecting difficulty, Random forest predicts the customer conditions. Since conditions for the customer are decided by in charge of bad debt collection through the negotiations with the customer, so that it is difficult to predict.

Table 10. Class Error by Customer Condition

Customer Conditions	Class Error
Claim difficulty	0.617187500
Collecting difficulty	0.001083907
Death with no inheritance	0.788461538
Missing	0.527859238
Personal Bankruptcy	0.486206897

Table11. indicates the mean decrease Gini by customer attributes. This figure means contribution of each parameter. The parameter of prefecture contributes characterize the customer next to the parameter of customer ID and Debt processing period.

Table 11. Mean Decrease Gini by Customer Attributes

Parameters	Mean Decrease Gini
Customer ID	651.91084
Prefecture	333.90053
Sex	59.35230
Age	203.55312
Debt Processing Period	651.05821
Payment Method	23.60707
Number of Payment	116.68290
Number of Remittance	213.72330
Phone Call Condition	136.34009
Number of Purchased Items	110.25428
Amount of Sales	307.83575

From the result of the analyses, we figured out the bad debt customers' characteristic that the region is an important factor, in addition to the long term evasion and big-ticket purchase.

5   Concluding Remarks

This paper presented investigating for analyzing customer characteristics from the bad debt list of a mail order company which aims to understand the characteristics of the bad debt customers.

We have described the research background, research method, and analytical results. So far, concerning screening for the potential bad debt customers, it depends on the heuristic knowledge based on the staffs' working experiences. In order to be expansion of the mail order industry, such a bad debt customer is also required. It is something of side-effect factor so far.

The analytical results have suggested that the one of the characteristics of the bad debt customers in the mail order industry, it turned out that prefecture is an important factor. Furthermore, the amount of money for order and ID are also important factor. These parameter mean that the bad debt customers order repeatedly and big-ticket. This result will make it use for the decision support knowledge for screening customer at the order received phase in the mail order industry.

Our future work includes; 1) Customer analysis with not only the black list but the white list of order to predict customer condition that fall into bad debt situation. 2) Comparison search regarding to the machine learning method to fit prediction of the customer characterization. 3) Finding new characteristics of the customer with another machine learning algorithms.

These works will require algorithm investigations and pay attention to hear the views of the rank-and-file employees for prediction and further survey studies.

We wish to express our great gratitude of the cooperation from the mail order company to our analysis.

References

1. The Japan Direct Marketing Association: 17th Annual National Mail Order Survey Report (2010)(Japanese)
2. The Nihon Keizai Shimbun (Nikkei): Nikkei Kaisha Jyoho, Summer (2012) (Japanese)
3. Simester, D.I., Sun, P., Tsitsiklis, J.N.: Dynamic Catalog Mailing Policies. Management Science 52(5), 683–696 (2006)
4. Kimijima, M.: A Study on Measuring Input-Output Process on Order-Getting Costs for Direct Marketing. Yokohama National University Departmental Bulletin Paper 16(1), 21–39 (2010) (Japanese)
5. Conlin, M., O'Donoghue, T., Vogelsang, T.J.: Projection Bias in Catalog Orders. American Economic Review 9(4), 1217–1249 (2007)
6. Matsuda, Y., Ebihara, J.: Forecasting Model in the Mail-Order Industry. UNISYS Technology Review 71, 52–68 (2001) (Japanese)
7. Motoda, H., Washio, T.: Perspective of Data Mining, System/Control/Information, the Institute of Systems. Control and Information Engineers 46(4), 169–176 (2002) (Japanese)
8. Ishigaki, T., Motomura, Y., Chan, H.: Consumer Behavior Modeling Based on Large Scale Data and Cognitive Structures. IEICE Technical Report, NC2008-157, 108(480), 319–324 (2009) (Japanese)
9. Yamashita S., Tsuruga T., Kawaguchi, N.: Consideration and Comparison about a Valuation Method of a Credit Risk Model, Discussion Paper Series (11), Financial Research Center, Financial Service Agency (2003) (Japanese)
10. Yano, J., Ikai, M., Nakagawa, K., Takahashi, S., Namatame, T.: A Study on Credit Card Users' Default Prediction. Journal of Operations Research of Japan 51(2), 104–110 (2006) (Japanese)
11. Sunayama, W., Yada, K.: Modeling and Analysis of Persuading Process using Conversation Logs. In: Proc. of the 20th Annual Conference of the Japanese Society for Artificial Intelligence, vol. 3C3-3 (2006) (Japanese)

12. Tanabe, K., Kurita, T., Nishida, K.: Prediction of Corporate Credit Ratings by Support Vector Machine. Journal of Japan Society for Management Information 20(1), 23–38 (2011) (Japanese)
13. Min, J.H., Lee, Y.C.: Bankruptcy Prediction Using Support Vector Machine with Optimal Choice of Kernel Function Parameters. Expert Systems with Applications 28(4), 603–614 (2005)
14. Abe, N., Melville, P., Pendus, C., Reddy, C.K., Jensen, D.L., Thomas, V.P., Bennett, J.J., Anderson, G.F., Cooley, B.R., Kowalczyk, M., Domick, M., Gardinier, T.: Optimizing Debt Collections Using Constrained Reinforcement Learning. In: Proc. of the 16th ACM SIGKDD International Conference on Knowledge Discovery and Data Mining, pp. 75–84. ACM (2010)
15. Hashimoto, M., Yoshida, K.: A Study on the Evaluation Technique of the Claim Assessment Model in a Retail Financial Sector. In: Proc. of The Annual Conference of The Japan Society for Management Information, pp. 63–66 (2004) (Japanese)
16. Breiman, L.: Random Forests. Machine Learning 45(1), 5–32 (2001)
17. Jin, M., Murakami, M.: Authorship Identification using Random Forests. Proc. of the Institute of Statistical Mathematics 55(2), 255–268 (2008) (Japanese)
18. Jin, M.: Estimation of When the Works were Written: With the Works of Ryunosuke Akutagawa as Examples. Behaviormetrics 36(2), 89–103 (2009) (Japanese)
19. Kobayashi, Y., Tanaka, S., Tomiura, Y.: Classification and Assessment of English Scientific Papers Using Random Forests. IPSJ SIG Technical Reports, 2011-CH, 90(6), 1–8 (2011) (Japanese)
20. Konishi, F., Uchida, S., Toda, K., Monden, A.: An Approach of Software Development Effort Estimation by Random Forests. In: Proc. of the 2011 IEICE General Conference, Information/System (1), vol. 17 (2011) (Japanese)
21. Ohashi, K., Toyoda, H., Kubo, S.: A Study on Classification and Prediction of the Potential Customers. Journal of Operations Research of Japan 56(2), 71–76 (2011) (Japanese)
22. Umezawa, Y., Mori, H.: Credit Risk Evaluation of Power Market Players with Random Forest. Trans. of the IEEJ, Power and Energy Society 128(1), 165–172 (2008) (Japanese)
23. The Statistics Bureau: Population Census (2010), http://www.stat.go.jp/english/index.htm

Author Index

Printed in the United States
By Bookmasters

Printed in the United States
By Bookmasters